In recent years there has been a host of new advances in our understanding of how we see. From molecular genetics come details of the photopigments and the molecular causes of disorders like colour blindness. In-depth analysis has shown how a cell converts light into a neural signal using the photopigments. Traditional techniques of microelectrode recording along with new techniques of functional imaging – such as PET scans – have made it possible to understand how visual information is processed in the brain. This processing results in the single coherent perception of the world we see in our 'mind's eye'. *An Introduction to the Visual System* provides a concise, but detailed, overview of this field. It is clearly written, and each chapter ends with a helpful 'key points' section. It is ideal for anyone studying visual perception, from the second year of an undergraduate course onwards.

An introduction to the visual system

An introduction to the visual system

Martin J. Tovée
Newcastle University, UK

Published by the Press Syndicate of the University of Cambridge

The Pitt Building, Trumpington Street, Cambridge CB2 1RP

40 West 20th Street, New York, NY 10011-4211 USA

10 Stamford Road, Oakleigh, Melbourne 3166, Australia

First published 1996

Printed in Great Britain at the University Press, Cambridge

A catalogue record for this book is available from the British Library

Library of Congress cataloguing in publication data available

ISBN 0 521 48290 9 hardback

ISBN 0 521 48339 5 paperback

SE

To my wife Esther and our daughter Charlotte

Contents

The Institute of Biology

The Institute of Biology is once again pleased to have joined with Cambridge University Press to publish a life science text. Many undergraduate biologists have found the Institute's *Studies in Biology* series (formerly published with Edward Arnold and now with Cambridge University Press) of great value; indeed, life science graduates have also found them useful update texts, so a number of these books have been revised and reprinted several times. However, the Institute's main membership consists of more senior biologists and it was felt that there was a need for larger, more advanced books. To this end, the Institute is delighted to be working with Cambridge University Press in producing biological texts for those who wish to build on the grounding that the *Studies in Biology* series provides.

Publishing is just one of the many activities carried out by the Institute of Biology, which is the professional body for life scientists charged by Royal Charter to represent UK biologists and biology (though increasingly the Institute is attracting overseas membership). The Institute has some 15000 members, it organises symposia both nationally and locally through its regional branch structure, provides careers literature for school leavers and undergraduates and actively comments on a range of science policy issues to industry, statutory bodies and governmental departments. It does not do this alone. For while the Institute of Biology is the body that represents the broad church of biology, there are some 75 specialist UK life science societies affiliated to it: altogether these bodies attract over 100000 subscriptions from the life science community, ranging from biochemists and molecular biologists through to whole-organism biologists, ecologists and those concerned with biosphere science. It is from this range that the Institute and Cambridge

University Press seeks to publish key texts. We hope you find them of value. If you have any comment to make on any of the joint texts or require further information about the Institute of Biology, then please write to the address below or find the Institute's pages on the InterNet.

The Institute of Biology
20–22 Queensberry Place
London SW7 2DZ

InterNet - **http://www.primex.co.uk/iob/iob.html**

(correct at date of going to press)

Acknowledgements

I am grateful for discussion of the topics covered in this book with a number of people. I should like to thank Malcolm Young for his comments on object recognition and particularly on the role of a temporal binding mechanism. I should also like to thank Edmund Rolls, Dave Perrett, Mike Oram, Roger Mason and Hugo Critchley for discussion of the issues involved in object and face recognition. I should also like to thank Sue Healy for discussion of, and her helpful comments on, the topic of learning and memory in the visual system. I am also grateful for past discussions of the principles of retinal colour vision and photopigment with John Mollon and Jim Bowmaker. Lastly, and most importantly, I would like to thank my wife and research collaborator Esther for her critical reading of, and comments on, the manuscript.

1

Introduction

A user's guide?

The aim of this book is to provide a concise but detailed account of how your visual system is organised and functions to produce visual perception. There have been a host of new advances in our understanding of how our visual system is organised. These new discoveries stretch from the structural basis of the visual pigments that capture light to the neural basis of higher visual function.

In the past few years, the application of the techniques of molecular genetics have allowed us to determine the genetic and structural basis of the molecules that make up the photopigments, and the faults that can arise and produce visual deficits such as colour blindness, night blindness and retinitis pigmentosa. Careful analysis has also allowed the changes in cell chemistry that convert the absorption of light by the photopigment into a neural signal to be understood. The use of functional imaging techniques, in concert with more traditional techniques such as microelectrode recording, have made it possible to understand how visual information is processed in the brain. This processing seems to be both parallel and hierarchical. Visual information is split into its different component parts, such as colour, motion, orientation, texture, shape and depth, and these are analysed in parallel in separate areas, each specialised for this particular visual feature. The processed information is then reassembled into a single coherent perception of our visual world in subsequent, higher visual areas. Recent advances have allowed us to identify which areas are performing these functions and how they interact with one another.

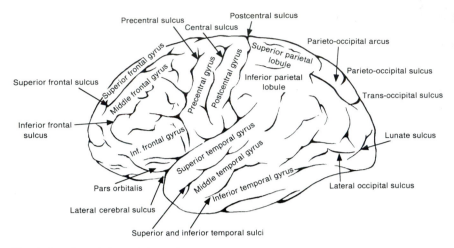

Fig. 1.1. Superiolateral view of the left hemisphere of the human cerebral cortex showing the names of the major gyri and sulci. (Redrawn from Bindman & Lippold, 1981.)

Many of the new advances have come from new experimental techniques such as magnetic resonance imaging (MRI) and positron emission tomography (PET), which allow direct, non-invasive measurement of how the human visual system functions. In this introductory chapter, I will first discuss the gross structure of the brain and then some of the new methods used to determine the function of different brain areas. To understand vision, we must understand its neural basis, and how this shapes and limits our perception.

Brain organisation

The mammalian cortex is a strip of neurons, usually divided into six layers. It varies in thickness from about 1.5 mm to 4.5 mm in humans, and this is not very different even for the very small cerebral hemispheres of the rat, where the thickness is about 1–2 mm. The most conspicuous difference is that the surface area increases enormously in higher animals. For example, the surface area is 3–5 cm^2 per hemisphere in small-brained rodents to 1100 cm^2 in humans. To accommodate this increase in surface area within the confines of the skull, the cortex is thrown into a series of ridges (*gyri*) and furrows (*sulci*) (see Fig. 1.1). In humans, about two thirds of the cortex is buried in the sulci. The cortex is divided into four main lobes; the occipital lobe, the temporal lobe, the parietal lobe and the frontal lobe. These lobes are then subdivided into different functional areas.

Looking at the brain in detail, we find that it is an incredibly complex structure. It contains around 10^{11} neurons, which have more than 10^{15} synapses and at least 2000 miles of axonal connections (Young & Scannell, 1993). Fortunately, for those of us who wish to make sense of how the brain works, there are several rules of organisation that simplify our task. First, neurons with similar patterns of connections and response properties are clustered together to form areas. For example, in the monkey and the cat there are about 70 cortical areas linked by around 1000 connections. Connections between these brain areas may consist of tens of thousands or even millions of nerve fibres. Many of these areas seem specialised to perform different tasks, so, for example, visual area 5 (V5) seems specialised to process information on visual motion and visual area 4 (V4) seems specialised for colour. The number of different specialised areas increases with increasing size and complexity of the brain. For example, mice have 15 cortical areas, of which around five are visual areas, whereas the cat has 65 cortical areas, of which 22 are visual (Kaas, 1989; Scannell, Blakemore & Young, 1995). It is suggested that the increase in visual areas allows the analysis of an increased number of visual parameters, which in turn allows a more complex and detailed analysis of visual stimuli. There is considerable interaction between neurons dealing with a particular visual parameter, such as colour or motion, and by grouping all such neurons into specialised areas the amount and the length of connections between neurons is reduced. The arrangement and connections between neurons is largely genetically pre-determined. To have widely interconnected neurons, and to have many different types of neuron with different connection patterns spread throughout the brain, would be extremely difficult to programme genetically and would have a greater potential for errors (Kaas, 1989).

Second, many of these different areas themselves are subdivided into smaller processing units. For example, in the primary visual area (V1), the cells are organised into columns within which all the cells have similar response properties. This form of columnar organisation seems to be a common feature within the visual system. Third, a further feature of organisation of the visual system is lateralisation. On either side of the brain, there is a duplication of visual areas. So there are two V1 areas and two V5 areas and so on. However, the higher visual areas, such as the inferior temporal cortex in monkeys and the inferior temporal and fusiform gyri in humans, do slightly different tasks on different sides of the brain. So, for example, the recognition of faces is mediated by the right side of the brain. This process of lateralisation allows the brain to carry out a greater variety of tasks with a limited amount of brain tissue.

Humans and Old World primates seem to have a visual system based on a broadly similar organisation. Differences seem to arise between the human and Old World monkey visual systems largely because of the expansion of the cortex in humans, which displaces the higher areas relative to their position in Old World primates. For this reason, during the course of this book, I will refer to the visual areas by the names originally coined for areas in the monkey cortex, but which are now being applied to human visual areas (see Fig. 1.2) (Kaas, 1992; Tootell *et al.*, 1995b). A problem with coming to grips with the visual system is that different research groups have used different names for the same area. For example, visual area 1 (V1) is also called the primary visual cortex and the striate cortex, and the higher visual areas can be collectively referred to as either the prestriate cortex or the extrastriate cortex. When I come to describe each area, I will use its most common name, but I will also list the other names by which you might encounter the area in other accounts of visual function.

Analysis techniques

Traditional methods of divining the function of brain areas have relied on two lines of approach: the study of human patients who have suffered brain damage and the use of animal models of human brain function. Common causes of head injuries to human patients are strokes, traumatic head injuries, such as those suffered in car accidents, and carbon monoxide poisoning. The difficulty with this approach is that the damage tends to be widespread, affecting more than one type of visual process. For example, damage that causes visual agnosia (the inability to recognise objects) is often linked to achromatopsia (an impairment of colour perception). The alternative line of investigation has been to use an animal model of human visual function. The advantage of this approach is that artificially induced lesions can be used to remove selectively specific brain areas, to determine their function. Also the activity of single neurons can be determined using a technique called microelectrode or single-unit recording. In this technique, a glass-insulated, tungsten-wire microelectrode is inserted into an animal's brain and its position adjusted until it is adjacent to a neuron in a particular brain area. The microelectrode can detect the small electrical changes associated with an action potential, and so the activity of single neurons in response to different visual stimuli can be determined.

Three new non-invasive analysis techniques have recently been developed to examine brain function: computerised tomography, magnetic resonance imaging and positron emission tomography. *Computerised tomography (CT)*, or

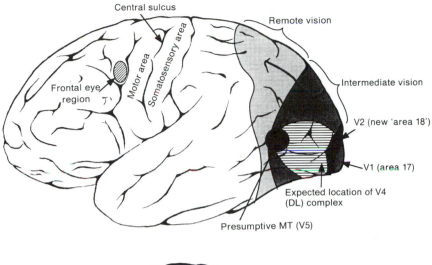

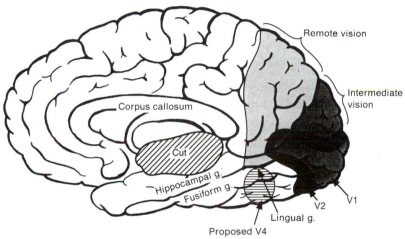

Medial view

Fig. 1.2. The putative location of some of the important visual functions in the human visual cortex, shown both in lateral view and in medial-cross section. V1, the primary visual cortex, also called the striate cortex; V2, visual area 2; V4, visual area 4, also called the dorsolateral (DL) complex in New World primates; MT, middle temporal, also called visual area 5 (V5); g. gyrus. (Redrawn and modified from Kaas, 1992.)

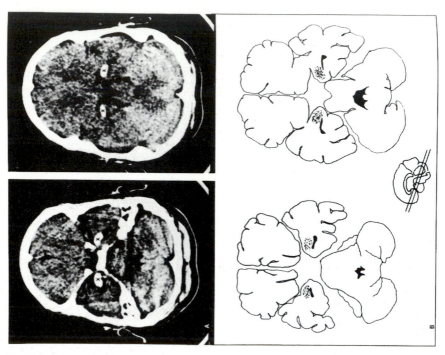

Fig. 1.3. Transverse CT scans of a women patient (S. M.) with Urbach–Wiethe's disease. In this condition, deposits of calcium are laid down in a brain area called the amygdala (indicated by X marks on the figure). The destruction of the amygdala disrupts the interpretation of facial expression (see Chapter 9). (Reproduced with permission from Tranel & Hyman, 1990. Copyright (1990) American Medical Association.)

computer-assisted tomography (*CAT*), uses X-rays for a non-invasive analysis of the brain. The patient's head is placed in a large doughnut-shaped ring. The ring contains an X-ray tube and, directly opposite to it on the other side of the patient's head, an X-ray detector. The X-ray beam passes through the patient's head, and the radioactivity that is able to pass through it is measured by the detector. The X-ray emitter and detector scan the head from front to back. They are then moved around the ring by a few degrees and the transmission of radioactivity is measured again. The process is repeated until the brain has been scanned from all angles. The computer takes the information and plots a two-dimensional picture of a horizontal section of the brain (see Fig. 1.3). The patient's head is then moved up or down through the ring, and the scan is taken of another section of the brain.

A more detailed picture is available from *magnetic resonance imaging* (*MRI*). It resembles the CT scanner, but instead of using X-rays it passes an extremely

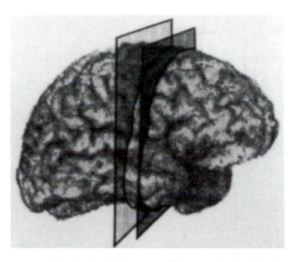

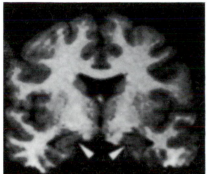 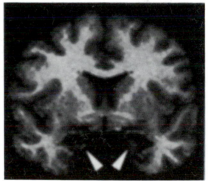

Fig. 1.4. An MRI scan of the same patient's brain. The planes of section are shown at the top on a three-dimensional reconstruction of S. M.'s brain. The lower right image shows that there is extensive bilateral damage to the amygdala, as indicated by the large arrow heads. The lower left image illustrates that this damage is restricted to the amygdala, and the neocortex and hippocampus are spared, as indicated by the small arrows. (Reproduced with permission from Adolphs *et al.*, 1994. Copyright (1994) MacMillan Magazines Ltd.)

strong magnetic field through the patient's head. When a person's head is placed in a strong magnetic field, the nuclei of some molecules in the body spin with a certain orientation. If a radio frequency wave is then passed through the body, these nuclei emit radiowaves of their own. Different molecules emit energy at different frequencies. The MRI scanner is tuned to detect the radiation from hydrogen molecules. Because these molecules are present in different concentrations in different brain tissues, the scanner can use the information to prepare pictures of slices of the brain (see Fig. 1.4). Unlike CT scans, which are limited to the horizontal plane, MRI scans can be taken in the sagittal or frontal planes as well.

These two techniques provide a representation of brain structure but do not provide any information on how the different parts of the brain function. A method that measures brain function, rather than brain structure, is *positron emission tomography* (*PET*). PET measurements depend on the assumption that active areas of the brain have a higher blood flow than inactive areas. This is because these more active areas use more oxygen and metabolites and produce more waste products. So an increased blood flow is necessary to supply the former and remove the latter. A PET camera consists of a doughnut-shaped set of radiation detectors that circles the subject's head. After the subject is positioned within the machine, the experimenter injects a small amount of water labelled with the positron-emitting radioactive isotope oxygen-15 (^{15}O) into a vein in the subject's arm. Over the minute following the injection, the radioactive water accumulates in the brain in direct proportion to the local blood flow. The greater the blood flow, the greater the radiation counts recorded by PET. The measurement of blood flow with ^{15}O takes about 1 min. The half life of ^{15}O is only 2 min, which is important as one does not want to inject long-lasting radioactive material into someone.

Different human brains vary slightly in their relative sizes and shape, and as PET scans do not provide any structural information, they are usually combined with MRI scans to allow the accurate comparison of structural and functional information (e.g. Zeki *et al.*, 1991). Although PET scanning is able to determine roughly which areas are active, its ability accurately to resolve specific regions is limited. A new technique that is now coming into use is *functional MRI* (*fMRI*) and this has better resolution. This method is a refinement of the MRI technique and, like PET scanning, it measures regional blood flow (Tanaka, Ogawa & Urgubil, 1992). Deoxyhaemaglobin (haemaglobin without a bound oxygen) is paramagnetic and so a blood vessel containing deoxyhaemaglobin placed in a magnetic field alters that field in its locality. It is, therefore, possible to map blood flow based on these changes on local magnetic fields.

These new techniques have allowed us to match behaviour to the anatomy and function of the brain. For example, when we perceive colour, we can now say which brain areas seem to be processing this information to give the sensation of colour. We can also see how different brain areas interact to produce the complex synthesis of different visual sensations that is our everyday experience of the visual world.

Key points

1. The human cortex is a strip of neurons, usually divided into six layers, that varies in thickness from about 1.5 to 4.5 mm. To fit it all into the confines of the skull, the cortex is thrown into a series of ridges (gyri) and furrows (sulci).

2. The cortex is divided into four main lobes: the occipital lobe, the temporal lobe, the parietal lobe and the frontal lobe. These lobes are then subdivided into different functional areas.

3. There are several rules of organisation for the brain. First, neurons with similar patterns of connections and response properties are clustered together to form areas. Second, these different areas themselves are subdivided into smaller processing units. Third, corresponding higher visual areas on different sides of the brain do slightly different jobs, a process called lateralisation.

4. Computerised tomography (CT), or computer-assisted tomography (CAT), uses X-rays for a non-invasive analysis of the brain. The X-ray emitter and detector scan the head from front to back, and the process is repeated until the brain has been scanned from all angles. The computer takes the information and plots a two-dimensional picture of a horizontal section of the brain. The patient's head is then moved up or down, and the scan is taken of another section of the brain.

5. Magnetic resonance imaging (MRI) passes an extremely strong magnetic field through the patient's head; this causes the nuclei of some molecules to emit radiowaves. Different molecules emit energy at different frequencies. The MRI scanner is tuned to detect the radiation from hydrogen molecules. Because these molecules are present in different concentrations in different brain tissues, the scanner can use the information to prepare pictures of slices of the brain.

6. Positron emission tomography (PET) measures the flow of radioactively labelled blood to different areas of the brain. It is assumed that an increased flow to a brain area is an indication of increased function. A more accurate measure of blood flow is functional magnetic resonance imaging (fMRI), which is a refinement of the MRI technique that maps blood flow based on changes in local magnetic fields.

2

The eye and forming the image

What is the eye for?

In this chapter, we will review the purpose of the eye and how the complex optical and neural machinery within it functions to perform this task. The basic function of the eye is to catch and focus light onto a thin layer of specially adapted sensory receptor cells that line the back of the eye. The eyeball is connected to an elaborate arrangement of muscles that allow it to move to follow target stimuli in the environment. The lens within the eye, which helps focus light, is also connected to muscles that can alter the lens' shape and, thus, its focal length. This allows target stimuli at different distances to be focused on the back of the eye. At the back of the eye, light energy is transformed into a neural signal by specialised receptor cells. This signal is modified in the retina, to emphasise changes and discontinuities in illumination, before the signal travels on to the brain via the optic nerve. In the sections that follow, we will examine these procedures in detail.

Light

Light has a dual nature, being considered both an *electromagnetic wave*, which can vary in frequency and wavelength, and also a series of discrete packets of energy, called *photons*. These forms of description are both used in explaining how the visual system responds to light. In determining the sensitivity of the visual system to light, such as the minimum threshold of light detection, it is usual to refer to light in terms of photons. However, when discussing colour perception, it is normal to refer to light in terms of its wavelength, measured in nanometres (nm). One nanometre is 10^{-9} m. For example, blue light is of comparatively short wavelengths (around 430–460 nm), whereas red light is of comparatively long wavelengths (around 560–580 nm).

Only electromagnetic radiation with a wavelength between 380 and 700 nm is visible to the human eye (Fig. 2.1). The width of the spectrum is determined primarily by the spectral absorbance of the photopigments in the eye. However, other structures play a role. Light just below the human visible spectrum (300–400 nm) is called *ultra-violet* (*UV*). The human lens and cornea absorbs strongly in this region, preventing UV light from reaching the retina (e.g. van den Berg & Tan, 1994). However, the human short-wavelength (or blue) photopigment's absorption spectrum extends into the UV range, and if the lens is removed, such as in cataract surgery, a subject can perceive UV light. A good reason for preventing UV light reaching the retina is that it is absorbed by many organic molecules, including DNA. Therefore, UV light, even of comparatively long wavelengths such as 380 nm, can cause retinal damage and cancer (van Norren & Schelkens, 1990). However, a wide variety of animal species show sensitivity to UV light, ranging from insects to mammals (Tovée, 1995a). Some have developed specific UV-sensitive photoreceptors to detect UV light, whereas others have combined a clear ocular media with short-wavelength receptors whose spectral absorbance extends into the UV range. These species use UV light for a wide range of purposes, from navigation using the pattern of UV light in the sky to intra specific communication using complex UV-reflecting patterns on their bodies.

The structure of the eye

The eyes are suspended in the orbits of the skull, and each is moved by six extra ocular muscles attached to the tough, fibrous outer-coating of the eye (the *sclera*). Within the orbit, the eye is cushioned by heavy deposits of fat surrounding the eyeball. The eyelids, movable folds of tissue, also protect the eye. Rapid closing of the eyelids (*blinking*) can occur both voluntarily and involuntarily. Blinks clean and moisten the surface of the eye; under normal circumstances we automatically blink about once every 4 s. It takes about a third of a second from the beginning of a blink, when the lids first begin to move, until they return to their resting point. For about half of this time, the eyelids are completely closed, reducing the amount of light reaching the retina by around 90%. If an external light is flicked on and off for this length of time, a brief blackout is very noticeable. So why do we not notice our blinks? One

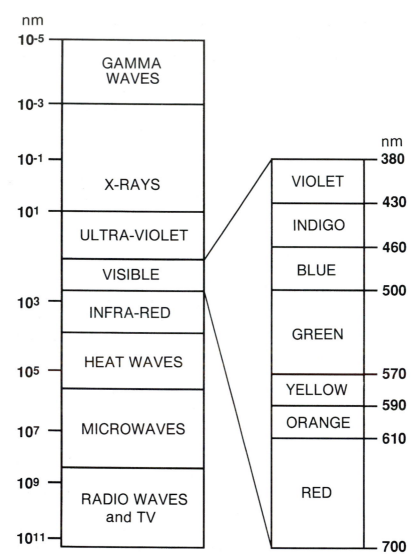

Fig 2.1. Light is a narrow band in the spectrum of electromagnetic radiation. Only electromagnetic radiation with a wavelength between 380 and 700 nm is visible to the human eye. The spectral sensitivity curves of the human eye is dependent upon the nature of the cone pigments. Other species have different cone pigments and can detect different ranges of electromagnetic radiation. For example, some birds seem to have five cone pigments, including one that absorbs in the ultra-violet. The brightly coloured plumage of birds visible to us is only a fraction of the patterns and colours birds can see. All non-primate mammals, including cats, dogs and horses, have only two cone pigments in their retinae and have poorer colour vision than humans. (Redrawn from Bruce and Young, 1990.)

suggestion has been that visual perception is suppressed during a blink. Evidence for this suggestion comes from an ingenious experiment by Volkman and his colleagues. The eyes lie directly above the roof of the mouth, and a bright light shone on the roof will stimulate the retina whether or not the eyes are closed. Volkman found that the light intensity required to stimulate the retina during a blink is five times greater than at any other time, strongly suggesting that there is suppression of perception during blinking (Volkman, Riggs & Moore, 1980).

A suppression of visual sensitivity during blinks explains why darkening is not seen, but it is not sufficient to account for the continuity of visual perception. Functional imaging has shown that in humans, just after a blink, the posterior parietal cortex is active (Harl, Salmelin & Tissari, 1994). The latency of the parietal activity suggests that it is a reaction to the eyeblink and does not occur in advance, as might be expected if it was connected with the generation of a motor command involved in the movement of the eyelids. This parietal activity is not seen if the blinks occur in darkness. The posterior parietal cortex is reciprocally connected with prefrontal cortical areas that seem to underlie spatial working memory, and it has been suggested that the parietal cortex is continually updated on information about the nature and structure of objects in a person's surroundings (Goodale & Milner, 1992) and is also kept informed about eyeblinks (Harl *et al.*, 1994). It is believed that this activity in the posterior parietal cortex is important for maintaining the illusion of a continuous image of the environment during each blink, perhaps by filling in the blink with visual sensation from working memory.

A mucous membrane, called the *conjunctiva*, lines the eyelid and folds back to attach to the eye (Fig. 2.2). The eye itself is a roughly spherical ball, about 2.5 cm in diameter. The sclera of the eye is made up of closely interwoven fibres, which appear white in colour. However, at the front of the eye, where the surface bulges out to form the *cornea*, the fibres of the sclera are arranged in a regular fashion. This part of the sclera is transparent and allows the entry of light. The part of the sclera surrounding the cornea is called the white of the eye. Behind the cornea is a ring of muscles called the *iris*. In the centre of the ring is an opening called the *pupil*, and the amount of light entering the eye is controlled by the pupil's diameter. The iris contains two bands of muscles, the dilator (whose contraction enlarges the pupil) and the sphincter (whose contraction reduces it). The sphincter is innervated by the parasympathetic nervous system, which uses the neurotransmitter acetylcholine. When we are interested in something, or someone, there is an unconscious expansion of the pupils. This is an important positive social signal. To mimic this response, and make themselves appear more attractive, women once added

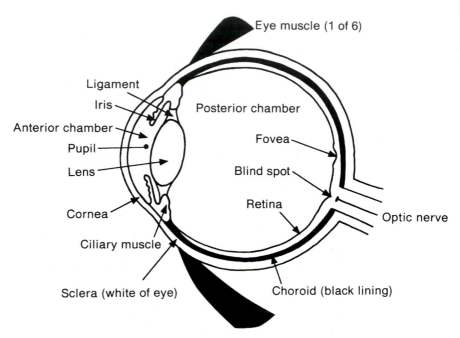

Fig. 2.2. Cross-sectional diagram of the human eye.

drops to their eyes containing the alkaloid atropine. This blocks the action of acetylcholine, causing dilation of the pupil by relaxing the sphincter of the iris. This preparation was made from deadly nightshade and gave this plant its species name of *Belladonna*, meaning beautiful women.

Beyond the pupil, the light passes through the *anterior chamber* of the eye to the lens. The anterior chamber is filled with a watery fluid called the *aqueous humour*. This fluid transports oxygen and nutrients to the structures it bathes and carries away their waste products. This function is normally carried out by blood in other parts of the body, but blood would interfere with the passage of light through the eye. The aqueous humour is being constantly produced by spongy tissue around the edge of the cornea (the *ciliary bodies*), and if the drainage is blocked or slowed, then pressure builds up in the eye. This can lead to permanent visual damage (*glaucoma*); this is one of the most common causes of blindness in Western Europe and North America.

The cornea and the lens alter the path of the light such that it will be in focus on the surface of the back of the eye, which is covered by the *retina*. The lens also inverts the image, so the picture of the world on the retinal surface is upside down. The inversion is not important as long as the relative spatial positions of the different features of the image are preserved. After passing

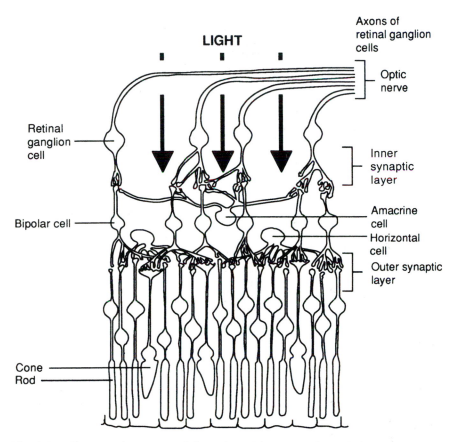

Fig. 2.3. The neural structure of the retina. Light passes through the neural layers before striking the receptors (rods and cones), which contain the photosensitive pigments. The vertical organisation of the retina is from receptor to bipolar cell to retinal ganglion cell. The horizontal organisation is mediated by horizontal cells at the receptor–bipolar (outer) synaptic layer and by the amacrine cells at the bipolar–retinal ganglion cell (inner) synaptic layer. (Redrawn from Cornsweet, 1970.)

through the lens, the light passes through the main part of the eye, which contains a clear, gelatinous substance (the *vitreous humour*), before reaching the retina. Unlike the aqueous humour, the vitreous humour is not being constantly replaced and so debris can accumulate. This debris can impinge on your visual awareness by forming floaters, small opacities that float about in the vitreous humour (White & Levatin, 1962).

The retina is divided into three main layers: the receptor cell layer, the bipolar layer and the ganglion cell layer (Fig. 2.3). The receptor cell layer is at the back of the retina and light has to pass thorough the transparent, overlying layers to get to it. The *photoreceptors* form synapses with *bipolar cells*, which in turn synapse onto *ganglion cells*, whose axons travel through the optic nerves to the brain. These axons come together and pass through the bipolar and receptor cell layer and leave the eye at a point called the *optic disk*. The optic disk produces a blindspot, as no photoreceptors can be located there. We are not consciously aware of the blindspot, as a result of a phenomenon called 'filling in' (Ramachandran, 1992). Based on the visual stimulus surrounding the blindspot, the visual system fills in the 'hole' in the visual image to give the complete picture of the world we are used to. This process seems to be mediated by the cortical neurons in areas V2 and V3 (de Weerd *et al.*, 1995). The retina also includes the outer plexiform layer, containing horizontal cells and the inner plexiform layer, containing amacrine cells. These cells transmit information in a direction parallel to the surface of the retina and so combine or subtract messages from adjacent photoreceptors.

Behind the retina is the *pigment epithelial layer*, and the ends of the photoreceptors are embedded in this layer. The specialised photoreceptors are unable to fulfil all their metabolic requirements and many of these, including visual pigment regeneration, are carried out by the pigment epithelial layer. Behind this layer is the *choroid layer*, which is rich in blood vessels. Both these layers contain the black pigment *melanin*; this light-absorbing pigment prevents the reflection of stray light within the globe of the eyeball. Without this pigment, light rays would be reflected in all directions within the eye and would cause diffuse illumination of the retina, rather than the contrast between light and dark spots required for the formation of precise images. Albinos lack melanin throughout their body and so have very poor visual acuity. Visual acuity is usually measured using an eye chart, such as the Snellen eye chart, which you are likely to see at any optician's. The measurement of acuity is scaled to a viewing distance of 20 feet (6 m) between the observer and the eye chart. Normal visual acuity is defined as 20/20. Someone with worse than normal visual acuity, for example 20/40, must view a display from a distance of 20 feet to see what a person with normal acuity can see at 40 feet (12 m). Someone with better than normal visual acuity, for example 20/10, can see from a distance of 20 feet what a normal person must view from 10 feet (3 m). Even with the best of optical correction, albino's rarely have better visual acuity than 20/100 or 20/200. For nocturnal or semi-nocturnal animals, like the cat, the opposite strategy is employed. Instead of a light-absorbent coating at the back of the eye, they have a shiney surface called a *tapetum*. This reflects light back into the eye, and although it degrades the resolution of the image, it increases the probability of a photon being absorbed by a photoreceptor. In low light intensity environments, this increases visual sensitivity; for a semi-nocturnal hunter this makes good sense. It also explains why the eye of a cat seems to glow when it catches the beam of a torch or any other light source.

Focusing the image

The ability of the eye to refract or focus light is primarily dependent on two structures: the cornea and the lens. When light passes from one medium to another of different density, the light's direction is altered: it is refracted. This happens when light passes from air into the cornea and from the aqueous humour in the eye to the lens. The relative difference in the density of the two sets of medium means that 70% of the eyes focusing is by the cornea. However, this focusing is not adjustable, whereas focusing by the lens is. The lens is situated immediately behind the iris, and its shape can be altered by the ciliary muscles. The lens is usually relatively flat, because of the tension of the elastic fibres that suspend it in the eye. In this state, the lens focuses distant objects on the retina. When the ciliary muscles contract, tension is taken off these elastic fibres, and the lens becomes more rounded in shape. In this condition, the lens focuses nearer objects on the retina. The ciliary muscles, thus, control whether near or far objects are focused, a process called *accommodation*. Accommodation is usually integrated with *convergence* ('turning together') of the eyes. When a near object is fixated, the eyes turn inward so that the two images of the object are fixated on corresponding portions of the retina.

The refractive or focusing power of the eye is measured in *diopters*, the reciprocal of the distance in metres between the eye and an object. For example, an eye with a refractive power of 10 diopters can bend light sufficiently to focus on an object 10 cm away. In humans with normal vision, the refractive power declines from 14 diopters at the age of ten (a focusing distance of only 7 cm, allowing one to focus on the tip of one's nose) to about 9 diopters at twenty (a focusing distance of 11 cm), 4 diopters in the mid-thirties (25 cm), 1–2 diopters in the mid-forties (50–100 cm) and close to zero by the age of seventy (a condition called *presbyopia*). The change from 4 to 2 diopters is the one people notice most as it affects reading. Most people hold books at 30–40 cm from their eyes. This change in focusing ability is related to changes in the lens's size, shape and flexibility (Koretz & Handelman, 1988).

The lens consists of three separate parts; an elastic covering (the capsule), an epithelial layer just inside the capsule and the lens itself. The lens is composed of *fibre cells* produced by the epithelial layer. The most common protein class in the lens is the *crystallins*. They make up 90% of the water-soluble proteins in the lenses of vertebrates. Most of the crystallins are contained in the

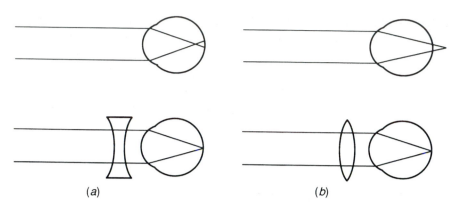

Fig. 2.4. (a) Myopic and (b) hyperopic eyes, showing the effects of optical correction. In the myopic eye, light rays from optical infinity are brought to a focus in front of the retina, which is corrected with a concave lens. The focal plane of the hyperopic eye lies behind the retina, and this can be corrected by a convex lens.

fibre cells. The unique spatial arrangement of these molecules is thought to be important for the maintenance of the transparency and refractive properties of the lens (Delaye & Tardieu, 1983). The distribution of the crystallins is not uniform throughout the lens. There is a general increase of protein concentration towards the centre of the lens. As a result, the refractive index increases towards the core of the lens, compensating for its changing curvature.

The fibre cells that make up the lens are being constantly produced, but none is discarded, which leads over time to a thickening of the lens. This cell production continues throughout life; as a result the lens gradually increases in diameter and slightly alters in shape. For example, the unaccommodated lens in an infant is 3.3 mm in thickness, whereas by the time a person reaches 70 years of age, the unaccommodated lens can be as thick as 5 mm. The old fibre cells in the centre of the lens become more densely packed, producing a hardening (sclerosis) of the lens. This thickening and hardening of the lens reduces its ability to focus light correctly on the retina. Moreover, those fibre cells in the centre eventually lose their nuclei and cell organelles. The crystallin in these cells cannot be replaced and, although it is a very stable protein, it does suffer over time a slight denaturisation (change in structure). This leads to a 'yellowing' of the lens, most noticeable in old age. This acts as a filter, subtly altering our colour perception as we grow older.

Two common problems arise with lens focusing: *myopia* and *hyperopia* (Fig. 2.4). Myopia (near sightedness) is an inability to see distant objects clearly. This problem can be caused in two ways: (a) refractive myopia, in which the cornea or lens bend the light too much, or (b) axial myopia, in which the eyeball is too long. As a result, the focus point is in front of the retina. Hyperopia (or far sightedness) is an inability to see nearby objects. In the hyperopic eye, the focus point is located behind the retina, because the eyeball is too short or because the lens is unable to fully focus the image (as discussed above). The changes in the lens means that hyperopia becomes more and more common with age. In contrast, myopia is more likely to develop in younger people (see below).

The development of myopia

Although myopia and hyperopia are relatively stable conditions in adults, in newborn infants these refractive errors rapidly diminish to produce *emmetropia* (this is when the length of the eye is correctly matched to the focal length of its optics). The young eye seems to be able to use visual information to determine whether to grow longer (towards myopia) or to reduce its growth and so cause a relative shortening of the eye (a change towards hyperopia). This process is called *emmetropisation*. In a ground-breaking experiment by Wiesel & Raviola in 1977, it was found that a degraded retinal image can lead to axial eye elongation, a condition called *deprivation myopia* or *form-deprivation myopia*. Further experiments have shown that if myopia or hyperopia is imposed by the use of spectacles on the young of a variety of species, including chicks, tree-shrews and primates, the shape of the developing eye alters to compensate for this change in focal length (Schaeffel, Glasser & Howland, 1988; Hung, Crawford & Smith, 1995).

It seems that one factor controlling eye growth is dependent on the local analysis of the retinal image without the necessity of communication with the brain. Severing the optic nerve does not alter the change in eye growth associated with deprivation myopia (Wildsoet & Wallman, 1992). The local retinal mechanism seems to be triggered by retinal image degradation, involving the loss of both contrast and high spatial frequencies.

Another factor in axial eye growth is the degree of accommodation the eye has to undergo to focus an image. This can be used as a measure of whether the eye is hyperopic (more accommodation) or myopic (less accommodation). However, chicks can still compensate for the addition of spectacle lenses after the ability of the eye to undergo accommodation has been eliminated by brain lesions or drugs (Schaeffel *et al.*, 1990). One reason for linking accommodation to myopia is that atropine (an antagonist for the muscarinic class of acetylcholine receptors), which blocks accommodation, has been said to halt the progression of myopia in children and monkeys (Raviola & Wiesel, 1985). It has been reported that treatment using atropine can produce a reduction in myopia of 1 diopter in children, suggesting that atropine reduces the progression of myopia (Wallman, 1994). Similarly, in children with one eye more myopic than the other, the difference can be reduced by treating the more myopic eye.

However, atropine may not act by blocking accommodation, because in chicks, which lack muscarinic receptors in their ciliary muscles, it reduces compensation for spectacle lenses (Stone *et al.*, 1988; Wallman, 1994). This suggests that atropine is working at the level of retinal muscarinic receptors; however, the levels required to produce myopia inhibition effects are far above the levels required to block muscarinic receptors, implying that non-specific drug effects or even retinal toxicity may be involved. Muscarinic blockers also reduce the synthesis of the sclera in chicks and rabbits, and so these blockers may be interfering with the normal, as well as the myopic, eye growth.

It has been suggested that there is a genetic component in the development of myopia, as children with two myopic parents are more likely to be myopic and have longer eyes than children with no myopic parents (Zadnik *et al.*, 1994). However, environment seems to have a stronger role. For example, myopia can be strongly correlated with education: 70–80% of Taiwanese students and Hong Kong medical students are myopic, compared with only 20–30% of the same age group in rural areas. Moreover, as once humans have become myopic they stay myopic, one can compare differences in occurrence at different ages in a population. In Finnish Inuits and in Hong Kong, the young are on average myopic whereas the middle-aged are not. This increase in myopia in the young suggests that environmental rather than hereditary factors are important in the development of myopia.

In conclusion, it seems that visual cues actively guide the growth of bird and mammal eyes towards emmetropia. This would be consistent with the association of education with myopia, as the students' eyes will grow into focus at the distance of the page, whereas those of a person who largely lives outdoors will grow to focus at infinity.

Clouding of the lens (cataracts)

Another important factor in focusing a clear image on the retina is the transparency of the lens. Clouding of the lens, which is called a *cataract*, is sometimes present at birth (a congenital cataract), can be caused by eye disease (a secondary cataract) or injury (a traumatic cataract), but the most common cause of all is old age (a senile cataract). Cataracts develop in roughly 75% of people over 65 and in 95% of people over 85. However, in only 15% of people do the cataracts cause serious visual impairment and in only 5% of patients is surgery necessary. In this case, a small opening is made in the eye, through which the lens is removed either by pushing on the lens to force it out and allowing its removal with forceps, or by a method called phacemulsification, which uses ultrasound to remove the lens. To compensate for removal of the lens, the patient may be given glasses, contact lens or intraocular lens (an artificial lens to replace the one removed).

Congenital cataracts may arise from a number of possible causes: aberrant function of the fibre cells, such as alteration of structural proteins or proteins that serve to protect the cell from damage and preserve the clarity of the lens matrix, or the defect may lie in a metabolic pathway, resulting in an accumulation and deposition of insoluble material in the lens. At least two forms of congenital cataract have been shown to be caused by mutations of the genes for the crystallin protein, which in turn lead to changes in the structure of this protein (Cartier *et al.*, 1994).

Photoreceptors

Once the image has been focused on the retina, this pattern of light must be transformed into a pattern of neural activity that can accurately represent the image. This transformation or *transduction* is carried out by the light-sensitive receptor cells (photoreceptors) in the retina. There are two types of photoreceptor: the *rods* and the *cones*. The human retina contains around 120 million rods and 6 million cones. The cones are concentrated in a small area of the retina called the fovea (Fig. 2.5). They mediate diurnal visual function and provide high–acuity colour vision. The rods mediate nocturnal vision and provide only low-acuity monochrome vision. There are three types or classes of cone. Between 5 and 10% of the total cone population are blue cones, and they form a ring or annulus around the edge of the fovea. The rest of the cones are red and green, in a ratio of 2:1. These last two classes do not seem to be arranged in a regular array but are randomly mixed together in small patches or clusters (Mollon & Bowmaker, 1992).

Photoreceptors consist of an outer segment connected by a cilium to an inner segment, containing the cell nucleus (Fig. 2.6). The outer segment contains several hundred thin membrane plates (*lamellae*); there are around 750 lamellae in a monkey rod. In rods, the lamellae are free floating disks, whereas in cones they consist of one continuous folded membrane. Embedded in the

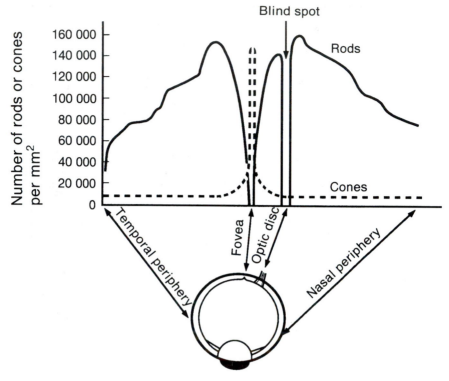

Fig. 2.5. The distribution of rods and cones over the extent of the retina of the right eye, as seen from above. In the left eye, the nasal and temporal areas of the retina would appear reversed, the relative distribution of rods and cones would be the same. The rods are completely absent from the centre of the fovea. (Redrawn with permission from Sekuler & Blake, 1994. Copyright (1994) McGraw-Hill.)

lamellae membrane are the photopigment molecules (*rhodopsin*). A single human rod contains 100 million photopigment molecules. They are so tightly packed together that there is only 20 nm between individual molecules, and they make up 80% of the membrane proteins. Each pigment molecule consists of two parts: *opsin* (a protein) connected by a Schiff-base linkage to *retinal* (a lipid), which is synthesised from retinol (vitamin A). Retinal is a long-chain molecule that can exist in two forms (or isomers), a straight-chain form (all-*trans* retinal) and a bent form (11-*cis*-retinal). 11-*cis*-Retinal is the only form that can bind to the opsin. When the 11-*cis* retinal absorbs a photon of light, the long chain straightens to the all-*trans* form, a process called *photoisomerisation*, and the photopigment molecule then eventually breaks into its two constituent parts. When this occurs, it changes colour from a rosy colour to a pale yellow. The photopigment is said to have been *bleached*.

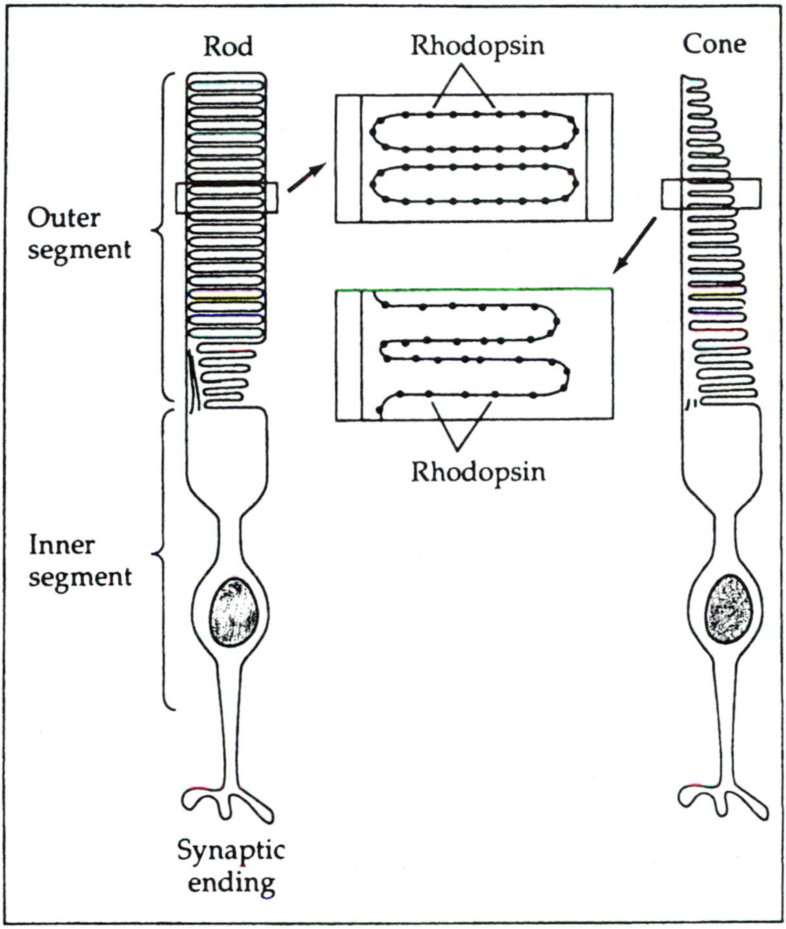

Fig. 2.6.   Schematic diagram of a rod and a cone. In the rod, the photopigment molecules (shown as black dots) are embedded in the membranes in the form of disks, not continuous with the outer membrane of the cell. In the cone, the molecules are on enfolded membranes that are continuous with the surface membrane. The outer segment is connected to the inner segment by a thin stalk. (Redrawn from Baylor, 1987.)

Transduction

In darkness, the rods and cones have a resting membrane potential of $-40\,$mV, considerably different from the usual membrane potential of $-80\,$mV found in other neurons. This is because a continuous *dark current* flows into the outer segment as sodium ions (Na^+) move down their electrochemical gradient through open cation channels. The effect of light is to cause *hyperpolarisation* of the cell membrane, by indirectly closing the cation channels in the outer segment membrane. This change in potential is in the opposite direction to the change found in other receptors and neurons, which depolarise when stimulated. The cation channels are normally kept open by cytoplasmic *cyclic guanosine 3′-5′-monophosphate (cGMP)*. The photoisomerisation of rhodopsin precipitates a series of reactions that result in a rapid reduction in the levels of cGMP. This, in turn, causes the cation channels to close and the electrical resistance of the outer segment membrane to increase, stopping the dark current. Therefore, the cGMP acts as an *internal messenger* within the cell, transferring news of the detection of light from rhodopsin molecules in the disk membrane to the ion channels in the cell membrane.

When a photon is absorbed by a rhodopsin molecule, the retinal chromophore undergoes photoisomerisation and changes from the 11-*cis* to the all-*trans* configuration, as the terminal chain connected to opsin rotates. This transition is very rapid, taking only $10^{-12}\,$s. The protein then goes through a series of intermediate forms. One of these intermediate forms is *metarhodopsin* II, which is produced 1 ms after the absorption of a photon. Metarhodopsin II is enzymatically active and binds to a disc membrane a globular or G-protein called *transducin*. This protein is composed of three subunits chains (T_α, T_β and T_γ). Different isoforms of the three subunits are present in rods and cones, and this may explain some of the physiological differences between the two receptor types (Peng *et al.*, 1992). In the inactive state, transducin is bound to a molecule of *guanosine diphosphate (GDP)*. Metarhodopsin II catalyses the exchange of a molecule of *guanosine triphosphate (GTP)* for the bound molecule of GDP. The metarhodopsin II–transducin–GTP complex then disassociates into metarhodopsin II, T_α-GTP and $T_{\beta\gamma}$. Metarhodopsin II can catalyse around 500 such exchanges before it is inactivated by phosphorylation of sequences near the C-terminus. This phosphorylation allows a protein called *arrestin* to compete with transducin for metarhodopsin II and so inhibits further catalytic activity.

Each of the released T_α-GTP molecules then binds to an enzyme called *phosphodiesterase (PDE)*, which is a complex of four subunits, two inhibitory subunits (PDE_γ) and two catalytic subunits (PDE_α and PDE_β). The interaction of T_α-GTP with PDE splits off the PDE_γ subunits, and the T_α-GTP·$PDE_{\alpha\beta}$ complex can then catalyse the break-up of cGMP. This break-up reaction produces one molecule of non-cyclic GMP and one H^+ for every molecule of cGMP hydrolysed. Around 800 molecules of cGMP are hydrolysed before the T_α component of the T_α-GTP·$PDE_{\alpha\beta}$ complex becomes deactivated. The deactivation occurs through the conversion of GTP to GDP, leading to the release of $PDE_{\alpha\beta}$; this then reassociates with the PDE_γ subunit. The T_α subunit, which is now again bound to GDP, reassociates with the $T_{\gamma\beta}$ subunits to complete the cycle initiated by the rhodopsin molecules absorption of a photon and storage of energy. This two-stage cycle is powered by the T_α-induced conversion of GTP to GDP. This system can allow the hydrolysis of 400 000 cGMP molecules within 1 s of the absorption of a single photon. The reduced levels of cGMP cause the sodium channels to close and the receptors to hyperpolarise. A single photon can close approximately 300 channels, about 3–5 % of the channels that are open in the dark. Internal levels of cGMP fall by about 20% on illumination (Baylor, 1987).

The calcium feedback mechanism

The intracellular concentration of calcium ions (Ca^{2+}) also changes over the course of the phototransduction process. In the dark, Ca^{2+} like Na^+ enters the cell through the open cation channels and is expelled from the cell by an electrogenic calcium–sodium exchanger, which is located in the cell membrane. The transduction process leads to a fall in intracellular cGMP concentration; the subsequent closure of the cation channels means that Ca^{2+} can no longer enter the cell, but the ions are still being pumped out of the cell. As a result, the levels of intracellular Ca^{2+} fall and do not start to rise again until the cation channels start to reopen as the cell recovers from stimulation. It has been shown that the changing level of Ca^{2+} acts as a feedback mechanism that speeds up a cell's recovery from light stimulation and also mediates light adaptation (Koutalos & Yau, 1993). Three mechanisms have been proposed for this action.

1. It has been suggested that intracellular Ca^{2+} alters the action of guanylate cyclase, the enzyme responsible for the synthesis of cGMP. This is mediated through a Ca^{2+}-binding protein called *recoverin*. It was suggested that this protein activated guanylate cyclase at low Ca^{2+} levels. Therefore, as the levels of intracellular Ca^{2+} fall, the activity of guanylate cyclase should increase. This raises the concentration of cGMP, allowing the cGMP-activated cation channels to reopen and the dark current to flow again. However, recent reports have contradicted these findings, and the role of recoverin is presently in some doubt (Hurley *et al.*, 1993).

2. The affinity of the cGMP-gated cation channels for cGMP appears to be decreased by Ca^{2+}. Therefore, as Ca^{2+} levels fall, the affinity of the channels for cGMP rises, which helps to offset the fall in cGMP levels.

3. The third component of the effect of Ca^{2+} is through *S-modulin*, a calcium-binding protein homologous to recoverin. This protein lengthens the lifetime of active PDE at high Ca^{2+} levels, and it inhibits phosphorylation of rhodopsin at the same Ca^{2+} levels (Kawamura, 1993). Therefore, one of the sites of Ca^{2+} modulation of the cGMP cascade is at the level of pigment inactivation.

Signal efficiency

It is common in textbooks to emphasise the stability of the rhodopsin molecule and imply that the photoreceptors are, therefore, extremely efficient at signalling the presence of light, as there is little or no background noise. Denis Baylor calculated that the spontaneous thermal isomerisation of a single rhodopsin molecule from the 11-*cis* to the all-*trans* form should occur about once every 3000 years, or 10^{23} times more slowly than photoisomerisation (Baylor, 1987). However, the retinal photoreceptors are actually very 'noisy'. They produce discrete electrical events in the dark that are indistinguishable from those evoked by light. This phenomenon limits visual sensitivity at low light levels, although recently it has been argued that under certain circumstances noise can play a constructive role in the detection of weak signals, via a mechanism known as *stochastic resonance* (Wiesenfeld & Moss, 1995). The random and spontaneous electrical events are strongly temperature dependent and have, therefore, been attributed to the thermal isomerisation of retinal; thus initiating the G-protein cascade that should signal the absorption of a photon. However, the thermal generation of dark events in photoreceptors requires activation energies in the range 23–27 kcal mol, which is significantly less than the energy barrier of 45 kcal mol required for photoisomerisation of retinal. Recent work has suggested that photoreceptor noise results from the thermal isomerisation of a small proportion of photoreceptor molecules (< 0.01%) in which the Schiff-base linkage between the retinal chromophore and the opsin is unprotonated (Barlow *et al.*, 1993). This deprotonation of the linkage destabilises the photopigment molecule by reducing the energy barrier for isomerisation to 23 kcal mol. Interestingly, the horseshoe crab (*Limulus*) has developed a method for reducing photoreceptor noise during the night, increasing the sensitivity of their eyes and so helping them locate a mate. At night, nerve signals from a circadian clock in the brain reduces the spontaneous activity in the photoreceptors by reducing the number of photopigment molecules in the unprotonated state. The mechanism for this change seems to be a lowering of the external pH in the vicinity of the photopigment-containing membrane (Barlow *et al.*, 1993).

The centre–surround organisation of the retina

There are around 126 million photoreceptors in the retina, each one signalling information on how much light is absorbed at a particular point in the retina. The information from the retina is transmitted to the brain via the axons of the ganglion cells, but there are only 1 million of these. Therefore, the retina has to condense and reorganise the information from the photoreceptors into a form that can be transmitted through the optic nerve. To consider how it does this, we must ask ourselves what is the basic purpose of a visual system? It is not just to signal the presence or absence of illumination; rather, it is to detect patterns of light from which information relating to the identity of objects and their spatial relationships in the environment can be derived. The first step in this process is the detection of differences in light at adjacent locations, which are likely to signal an edge or border. Such edges can then be used to build up a picture of the environment (see Chapter 8). Regions of uniform illumination are less important, as they are unlikely to signal an edge.

The first point at which information relating to regional light differences is extracted is at the level of the retinal ganglion cells. Each ganglion cell is connected to a number of photoreceptors via bipolar cells. Stimulation of the retinal area corresponding to these photoreceptors alters the activity of the ganglion cell; this retinal area is called the ganglion cell's *receptive field*. The photoreceptors in a particular receptive field do not simply stimulate the ganglion cell but instead are arranged in what is called a *centre–surround organisation*. For example, light falling on the centre of the receptive field might result in the corresponding photoreceptors stimulating the ganglion cell (an *ON* response). Light falling just on the surrounding ring of photoreceptors might inhibit the ganglion cells (an *OFF* response). This cell is an example of an ON-centre, OFF-surround cell. There are also cells with the opposite arrangement, an OFF-centre and an ON-surround. This opponent interaction is often called *lateral inhibition*.

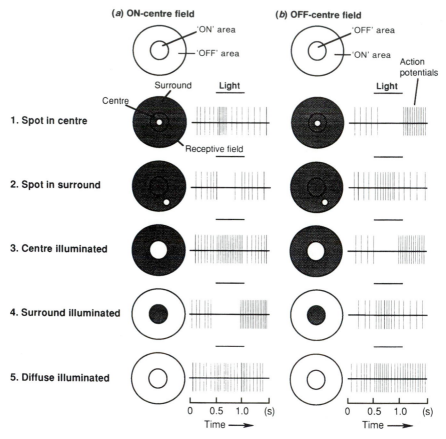

Fig. 2.7. Responses of retinal ganglion cells with ON-centre (a) and OFF-centre (b) receptive fields to various stimuli. (Redrawn from Kandel & Schwartz, 1982.)

If a whole receptive field is illuminated (as in Fig. 2.7a), both the ON-centre and OFF-surround are stimulated. The ON-centre excites the ganglion cell, but the OFF-surround inhibits it and there is little if any change in the ganglion cell's firing rate. Now consider what happens if an edge is positioned as shown in Fig. 2.7b. The ON-centre receives an increase in light and will stimulate the ganglion cell, while the OFF-surround receives a reduced level of light and will not inhibit the cell as much as before. The net result is that the ganglion cell is stimulated and will signal the presence of a light/dark boundary. Because the receptive fields of these centre–surround cells are usually concentrically arranged, the cells will respond well whatever the orientation of the edge. Only by comparing the responses of a number of ganglion cells can the orientation of a stimulus be determined, and this comparison is made in the first cortical visual area.

Table 2.1. *Range of visible light intensities*

	Intensity (candelas/m^2)	
The sun at noon	10^{10}	
	10^9	Damaging
	10^8	
	10^7	
Filament of a 100 Watt light bulb	10^6	
	10^5	
White paper in sunlight	10^4	Photopic vision
	10^3	
	10^2	
Comfortable reading	10	
	1	Mesopic vision
	10^{-1}	
White paper in moonlight	10^{-2}	
	10^{-3}	
White paper in starlight	10^{-4}	Scotopic vision
	10^{-5}	
Weakest visible light	10^{-6}	

Source: Taken from Sekuler and Blake, 1994.

Light adaptation

Humans can see over a range of illumination levels of about 10^{10}–10^{12}:1 (Table 2.1). However, all the information leaving the eye travels in the optic nerve fibres, whose response range is quite limited (perhaps 100:1), and so an enormous range of inputs is mapped onto a very small range of outputs. To cope with this problem, the visual system employs a number of strategies. First, the ganglion cell's response is dependent upon the average illumination of the retina. For example, if one examines the response of an ON-centre/OFF-surround cell, the strength of the response produced by the ON-centre response will be dependent upon the intensity of the illumination of the surround. The result is the shift of the ganglion cell's response function, as detailed in Fig. 2.8. This mobile response function has several important advantages over a fixed relationship. For a fixed relationship, the visual system will be insensitive to all but relatively large changes in input. However, the actual visual system can detect changes in illumination of less than 1%. Moreover, the fixed relationship is a wasteful system because at any one time the light intensities to which the visual system is exposed mostly lie within a small range. So, it is more effi-

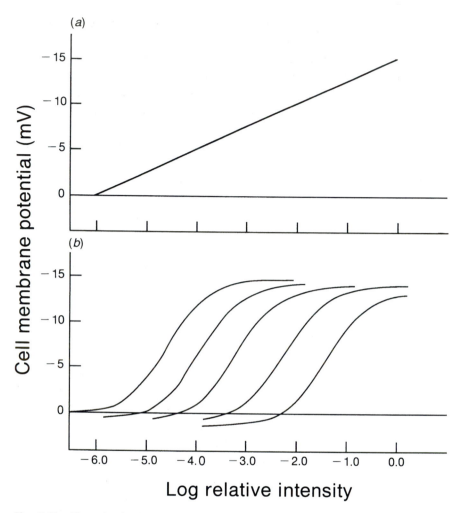

Fig. 2.8. The stimulus–response relationship for a cone, showing the amplitude of hyperpolarisation as a function of light intensity. (*a*) Without light adaptation. Hypothetical relationship shows how the whole range of working intensities is mapped onto the limited response range of a photoreceptor. This results in very poor sensitivity to small changes in illumination. (*b*) With light adaptation. The stimulus–response relationship is shown for a single cone in the turtle *Pseudemys scripta elegans,* as measured by Normann and Perlman (1979). The resting membrane potential in the dark is indicated by the line at 0 mV. Steady background light reduces the sensitivity of the cone but causes only small changes in the resting potential, permitting the cone to generate relatively large changes in membrane potential for relatively small increments or decrements in illumination about the background level. (Redrawn and modified from Lennie & D'Zmura, 1988.)

cient if the limited response range of the visual pathways are available to handle the whole of the relatively small range of light intensities likely to be encountered at any one time. It is more efficient if there is a flexible mapping of input onto output, with the operating range of the visual system shifting with the level of ambient light. This shift in the operating range and the accompanying change in sensitivity is called *light adaptation*. This shift in sensitivity allows a phenomenon called *lightness constancy*. A particular object or surface will appear equally light, relative to surrounding surfaces, over a range of illumination. Perceptual constancies, such as lightness, colour and object constancies, are a fundamental feature of the visual system. Features such as lightness, colour or shape can be used to identify an object and then can be used as a cue to subsequently recognise it. These perceptual cues would be useless if they changed with different viewing conditions, such as changes in illumination or viewing angle. As a result, the visual system has developed a number of mechanisms to allow the stimulus features of objects to appear constant under different conditions, and lightness constancy is one of these.

Duplicity theory of vision

Another strategy to expand the visual system's operating range is specialisation and division of labour. As mentioned above, the retina contains two types of photoreceptors, rods and cones, and the range of possible light intensities is divided between them. The rods respond to low intensities and the cones to high intensities (Fig. 2.9). There are three cone classes: red, green and blue (see Chapter 3); the photopic sensitivity shown in Fig. 2.9 is based on their combined activity. At intermediate intensity levels, there is a degree of overlap between the two photoreceptor systems, with both being active at the same time. Under these conditions, the rod system sums its responses with the responses of the red cone class. The spectral sensitivity of the rods and cones differs (see Fig. 3.1, p. 39) and when the rods and cones interact under these conditions, our perception of colour is shifted towards shorter wavelengths. This is called the *Purkinje shift*.

This division of labour between the two photoreceptor systems is termed the *duplicity theory* of vision and was first proposed by von Kries in 1896. The arrangement can be demonstrated in a number of ways. For example, the dark adaptation curve (the increase in sensitivity that occurs when illumination changes from light to dark) is clearly a two-stage function (Fig. 2.10). At high light intensities, the light-adapted eye is at its least sensitive. When the ambient light is turned off, then the sensitivity of the eye increases. It does so in two

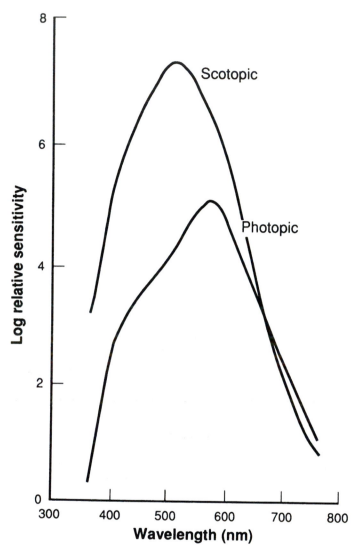

Fig. 2.9. Spectral sensitivity of the rods (scotopic curve) and cones (photopic curve). The rods can detect light at lower intensities than the cones, and their peak sensitivity is for lower wavelengths of light. (Redrawn from Kaufman, 1974.)

stages. It first increases for about 3–4 min, then levels off for about 7–10 min before increasing again for about 20–30 min. The first increase in sensitivity results from the cones, the second from the rods.

This difference in sensitivity is caused by the different rates of pigment regeneration in the rods and cones, which can be measured in the following way. A dim measuring beam of constant intensity is projected into the eye.

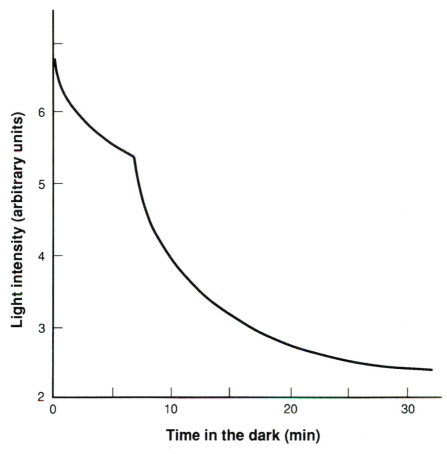

Fig. 2.10. This graph shows dark adaptation or the recovery from light adaptation. The curve represents the increasing sensitivity (or decreasing thresholds) for light detection with time in the dark. The initial rapid phase reflects the recovery of the cones and the slower phase that of the rods. An asymptote is reached after about 30 min in darkness.

The beam passes through the retina, hits the back of the eye and is reflected back out. During this procedure, much of the light is absorbed by the visual pigment, other structures in the eye and the black pigment epithelium. However, some of the light is reflected out. When the visual pigment is bleached it absorbs less light. Therefore, the amount of light that is reflected out of the eye will be a measure of the amount of pigment present. This method is called retinal densitometry and was used by William Rushton (1961) to measure the concentration of visual pigments during pigment regeneration. There are no rods in the centre of the fovea, and so if the measuring beam is confined to this region it is possible to measure the rate of pigment regeneration in cones: approximately 6 min to fully regenerate. To measure the rate of pigment regeneration in rods, Rushton used the technique on the retinae of rod monochromats (subjects who lack cones in their retinae) and found it took 30 min for rod pigment to regenerate.

The differences are also evident in the response of the eye to a flickering light. A brief flicker excites the eye for around 0.1–0.2 s, and because of the persistence of excitation, rapid successive flashes of light become fused together to give the appearance of being continuous. This phenomenon is used in films and television to give the illusion of a continuous image (the film screen flickers at 24 Hz and that of the television at 60 Hz). The frequency at which flicker fusion occurs, called the *critical frequency for fusion*, varies with light intensity. At low light intensity, flicker fusion can occur at 2–6 Hz, but under bright illumination, flicker fusion can be as high as at 60 Hz. The difference stems largely from the fact that the cones are active under bright illumination, and they have a shorter persistence in response to light. Under poor illumination, the rods mediate vision and they have a longer persistence. For the same reason, i.e. differences in persistence, the fusion frequency varies with retinal eccentricity: it is high at the fovea (dominated by cones) and lower in the periphery (dominated by rods).

Sensitivity, acuity and neural wiring

As mentioned above, there are over 120 million rods and at least 6 million cones but only 1 million ganglion cells to carry the information from the retina to the brain. This suggests a general convergence factor of 126:1, but in reality the convergence factor varies with retinal eccentricity. At the centre of the fovea, the convergence factor may be as low as 1:1, but in the periphery it may be as high as several hundred to one. The degree of convergence determines the spatial resolution of that part of the retina. The resolution (measured as *visual acuity*) refers to the ability to distinguish differences in the spatial distribution of light in the image. The size of a ganglion cell's receptive field will be large if the convergence factor is high, but receptive field size will be small if the convergence factor is low. The receptive fields can be thought of as analogous to the grain size in a photograph. The larger the grain size, the poorer the quality of the picture and the poorer the detail that can be discerned. At the centre of the retina (which is dominated by cones), the ganglion cell receptive fields are at their smallest and here spatial resolution is at its best. As one moves out into the periphery (where the rods dominate), the receptive field size rapidly increases and visual acuity rapidly decreases.

An increase in receptive field size does have one advantage; it increases sensitivity. If one considers a single ganglion cell, for it to be sufficiently stimulated to signal the presence of light, a certain amount of excitatory neurotransmitter must be released by the neurons that synapse onto it. If a dim light illuminates the retina, only a small proportion of photoreceptors will be stimulated. If the receptive field is small, with only a few photoreceptors, then perhaps only one of these photoreceptor will be stimulated. The amount of neurotransmitter released from that cell's excitation may not be enough to stimulate the ganglion cell. However, if the receptive field is large, the same proportion of photoreceptors will be stimulated, but a larger receptive field may contain several activated photoreceptors. The combined effects of these excited photoreceptors will stimulate the ganglion cell, where a single excited photoreceptor could not.

The rod system is more sensitive than the cone system for two additional reasons. First, on average, the rods have a larger diameter and are longer. The increased diameter increases the probability of a photon passing through an individual rod, and the increased length increases the probability of the absorption of the photons as they pass through the receptor. Second, the persistence of the response to the absorption of a photon is longer in rods than in cones. If a rod is stimulated by the absorption of a photon but this stimulation is too weak to stimulate the bipolar cell, then the probability of it absorbing a second while the rod is still excited is increased. This second absorption may boost the level of stimulation in the rod, such that it may be strong enough to initiate the passage of a signal in the bipolar and ganglion cells.

Key points

1. For humans, visible light is a narrow range of the electromagnetic spectrum of 380–700 nm. Many species can see shorter wavelengths (down to around 300 nm), a part of the spectrum called ultra-violet.
2. Light entering the eye is focused by the cornea and the lens. The cornea is responsible for 70% of the focusing in the eye, but its focal distance is not adjustable. The adjustable, 'fine-tuning' of the focusing of the image is carried out by the lens. The focusing power of the eye is measured in diopters, the reciprocal of the distance in metres between the eye and an object.
3. Two common problems arise with lens focusing: myopia and hyperopia. Myopia (near sightedness) is an inability to see distant objects clearly. Hyperopia (or far sightedness) is an inability to see nearby objects.

4. Although myopia and hyperopia are relatively stable conditions in adults, in newborn infants these refractive errors rapidly diminish to produce *emmetropia* (this is when the length of the eye is correctly matched to the focal length of its optics). The young eye seems to be able to use visual information to determine whether to grow longer (towards myopia) or to reduce its growth and so cause a relative shortening of the eye (a change towards hyperopia).

5. Once the image has been focused on the retina, this pattern of light must be transformed into a pattern of neural activity that can accurately represent the image. This transformation or transduction of light into neural energy is carried out by the light sensitive receptor cells (photoreceptors) in the retina. There are two types of photoreceptors: the rods and the cones.

6. Each photopigment molecule consists of two parts, opsin (a protein) connected by a Schiff-base linkage to retinal (a lipid), which is synthesised from retinol (vitamin A). Retinal is a long-chain molecule that can exist in two forms, a straight chain form (all-*trans* retinal) and a bent form (11-*cis* retinal). 11-*cis* Retinal is the only form which can bind to the opsin. When the 11-*cis* retinal absorbs a photon of light, the long chain straightens to the all-*trans* form, a process called photoisomerisation, and the photopigment molecule then eventually breaks into its two constituent parts.

7. In darkness, the rods and cones have a resting membrane potential of $-40\,\mathrm{mV}$, because a continuous dark current flows into the outer segment as Na^+ move through open sodium channels in the cell membrane. The effect of light is to cause hyperpolarisation of the cell membrane, by indirectly closing the cation channels in the outer segment membrane. The cation channels are normally kept open by cytoplasmic cGMP. The photoisomerisation of rhodopsin precipitates a series of reactions that result in a rapid reduction in the levels of cGMP. This in turn causes the cation channels to close and so reduces or stops the dark current.

8. The intracellular concentration of Ca^{2+} also changes over the course of the phototransduction process. The changing level of Ca^{2+} acts as a feedback mechanism that speeds up a cell's recovery from light stimulation and also mediates light adaptation.

9. Each retinal ganglion cell is connected to a number of photoreceptors via bipolar cells. Stimulation of the retinal area corresponding to these photoreceptors alters the activity of the ganglion cell; this retinal area is called the ganglion cell's receptive field. The photoreceptors in a

particular receptive field do not simply stimulate the ganglion cell but, instead, are arranged in what is called a centre–surround organisation.

10. The response range of the visual system is limited; however, the visual system responds to a wide range of light intensities, although we are likely to encounter only a relatively small range at any one time. There is a flexible mapping of input onto output, with the operating range of the visual system shifting with the level of ambient light. This shift in the operating range and the accompanying change in sensitivity is called light adaptation and it allows a phenomenon called lightness constancy. A particular object or surface will appear equally light, relative to surrounding surfaces, over a range of illumination.

11. To further expand the visual system's operating range, the rods and cones function over different ranges of light intensity. The rods respond to low intensities and the cones to high intensities. At intermediate intensity levels, there is a degree of overlap between the two photoreceptor systems. This arrangement is termed duplicity.

12. There are over 120 million rods and at least 6 million cones, but only 1 million ganglion cells to carry the information from the retina to the brain. At the centre of the fovea, the convergence factor may be as low as 1:1, but in the periphery it may be as high as several hundred to one. The degree of convergence determines the spatial resolution and sensitivity of that part of the retina. A low convergence allows good spatial resolution, but a low light sensitivity. A high convergence means poor spatial resolution but a higher light sensitivity.

3

Retinal colour vision

Why do we need more than one cone pigment?

In the vertebrate eye, colour is detected by cone receptors. In the case of humans and Old World primates, there are three cone classes (Fig. 3.1 and Table 3.1). A *blue* or short-wavelength pigment absorbing maximally at 420 nm, a *green* or middle-wavelength pigment absorbing maximally at 530 nm and a *red* or long-wavelength pigment absorbing at 565 nm (Dartnall Bowmaker & Mollon, 1983). For an animal to be able to discriminate between colours, it must have two or more different classes of cone. This is because a single cone pigment cannot discriminate between changes in wavelength and changes in the intensity of a light. For example, a red cone will respond strongly to a 560 nm light, but weakly to a 500 nm light. However, the same pattern of response can be obtained by a light of fixed wavelength, say 560 nm, and changing intensity, as a single cone class can only signal the number of photons absorbed by its pigment. This pattern of response is called *univariance*. To make the crucial differentiation between wavelength and intensity, a comparison of signals from two or more cone classes is required: 540 nm and 640 nm lights will produce different patterns of firing in the red and green cones compared with two 540 nm lights of different intensity. As a general rule of thumb, the more cone classes in an eye, the better will be the wavelength discrimination. Non-primate mammals, which rely heavily on sound and smell, have only two pigments (*dichromacy*), whereas birds, such as the pigeon, who are highly visually oriented have five (*pentachromacy*).

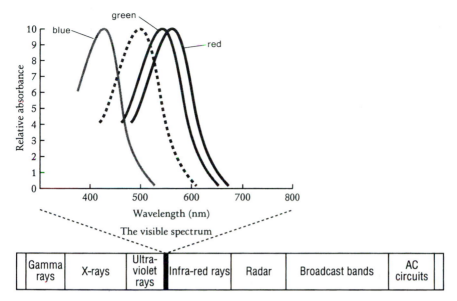

Fig. 3.1.   Spectral absorbance curves for the human photopigments. The three cones are shown in solid lines: a blue or short-wavelength pigment absorbing maximally at 420 nm, a green or middle-wavelength pigment absorbing maximally at 530 nm and a red or long-wavelength pigment absorbing at 565 nm. The rod pigment is indicated by a dotted line and absorbs maximally at 499 nm. (Redrawn and modified from Dartnall *et al.*, 1983.)

Trichromacy

In 1802, Thomas Young correctly proposed that the human eye detected different colours because it contained three types of receptor, each sensitive to a particular hue. His theory was referred to as the *trichromatic* (three colour) theory. It suggested that for a human observer any colour could be reproduced by various quantities of three colours selected from various points in the spectrum, such as red, green and blue. However, it has long been thought that there are four primary colours: red, green, blue and yellow. The trichromatic theory cannot explain why yellow is included in this group. In addition, some colours appear to blend while others do not. For example, one can speak of a bluish-green or a yellowish-green, but one cannot imagine a greenish-red or a bluish-yellow. To account for this, an alternative theory was proposed by Ewald Hering; this regarded mechanisms in the eye sensitive to these colours as undergoing some form of *opponent* interaction. As we discussed in the preceding paragraph, the responses from the three different

Table 3.1. *Primate visual pigments: the peak spectral sensitivities*

	Peak spectral absorbance (nm)			
	Blue cones	Rods	Green–red cones[a]	
Old World primates				
Grivet monkey (*Cercopithecus aethiops*)	434	499	535	566
Diana monkey (*C. diana*)	432	496	531	566
Spot-nosed monkey (*C. petaurista*)	424	497	534	563
Moustached guenon (*C. cephus*)	432	498	533	565
Talapoin (*C. talapoin*)	429	495	533	564
Patas monkey (*Erythrocebus patas*)	432	499	533	566
Rhesus monkey (*Macacamulatta*)	430	501	536	566
Cynomologus macaque (*M. fascicularis*)	429	502	532	567
Baboon (*Papio papio*)	426	500	536	566
Human (*Homo sapiens*)	420	499	530	565
Gorilla[b] (*Gorilla gorilla*)	—[c]	—	530	565
Chimpanzee (*Pan troglodytes*)	—	—	530	565
Orangutan (*Pongo pygmaeus*)	—	—	530	565
New World primates				
Squirrel monkey (*Saimiri sciureus*)	431	499	538	551 561
Dusky titi (*Cellibus moloch*)	—	—	538	549 561
Tufted capuchin (*Cebus apella*)	—	499	534	550 562
Spider monkey (*Ateles geoffroyi*)	430	—	—	550 563

Table 3.1. (*cont.*)

	Peak spectral absorbance (nm)				
	Blue cones	Rods	Green–red cones[a]		
Common marmoset	423	499	543	556	563
(*Callithrix jacchus*)					
Saddle-backed tamarin	436	—	545	557	562
(*Saguinus fuscicollis*)					
Owl monkey		505	543		
(*Aotus trivirgatus*)					
Prosimians					
Tree shrew	444	496	555		
(*Tupaia glis*)					
Black lemur	—	501	543		
(*Lemur macaco*)					
Bushbaby	—	501	552		
(*Galago crassicaudatus*)					

Notes:
[a] Old World primates have two cone classes in the red–green range, Prosimians have only one and most species of New World monkeys have three. However, individual males will have only one of these cone classes and a female will have either one or two of these cone classes.
[b] In these species, the peak spectral sensitivities are inferred from the gene structure and/or behavioural experiments.
[c] A dash indicates that no information on this particular pigment in this species has yet been obtained.

cone classes are compared to allow colour discrimination. This is indeed done in an opponent manner. There are three opponent mechanisms (Fig. 3.2). The first compares the difference between the red and the green cone classes. The second compares the difference between the blue cones and the sum of the red and green cones (yellow). The final mechanism is an achromatic (black–white) mechanism detecting differences in luminance. So, the human visual system is trichromatic and also compares the four primary colours in an opponent mechanism.

The basis of the opponent mechanism is the centre–surround opponency described in Chapter 2. The difference in illumination between the centre and the surround of ganglion cell receptive fields can obviously provide the basis

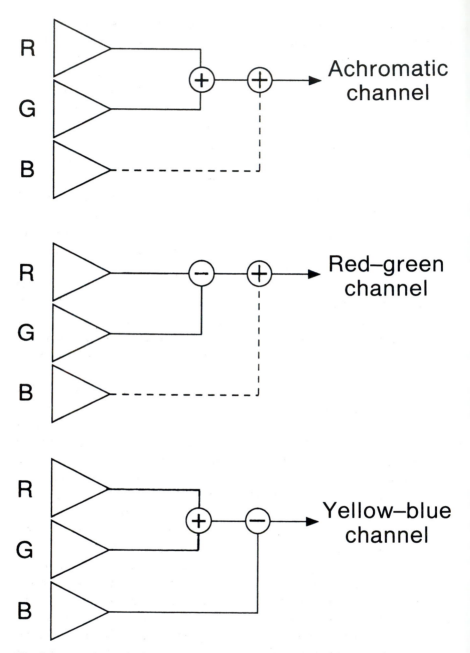

Fig. 3.2. A schematic diagram illustrating the relationship of the three opponent mechanisms. In the achromatic channel and the red–green channel, the role of the blue cones is unclear. They may under some conditions contribute to both achromatic and red–green sensitivity.

of the achromatic mechanism. But this arrangement can also provide colour information. If the centre is composed of cones of one class, and the surround of cones of another class, then centre–surround opponency can produce colour opponency. This idea suggests that the cones should be organised in the fovea in a regular pre-ordained array. However, experimental evidence suggests that the red and green cones are randomly arranged in the fovea (Mollon & Bowmaker, 1992). In the fovea, the receptive fields are very small and comprise a comparatively small number of cones. Given a random arrangement of the two cone classes, the probability is that there will be an unequal number of the two cone classes in the centre and the surround. Thus, there will be colour opponency without the need for a regular arrangement of cones. This almost haphazard arrangement emphasises the recent nature of the red and green cones in the eye of the Old World primate. All other mammals are dichromats, possessing a blue cone class and a second cone class that absorbs primarily in the red–green range. As in primates, the few blue cones form a ring around the edge of the fovea, and the rest of the fovea is composed of a single cone class. The centre–surround opponency in this part of the fovea merely signals achromatic differences. The recent addition of a third cone class (less than 35 million years ago) has been overlaid onto this older achromatic opponent system to produce a new colour opponent system (Mollon, 1989).

As can be seen from Fig. 3.1, the red and green pigments have a great deal of overlap in their spectral absorbance curves. There is a separation of only around 35 nm between their maximum spectral absorbances. Although colour discrimination relies on there being some overlap between cone pigments to allow comparison of the responses of two or more cone classes to a light, human colour vision would be improved by a more equal spacing of the three cone pigments in the spectrum. The reason that the red and the green pigments are not further apart may be linked to the way information from the cones is processed. As well as using the differences in the responses of different cone classes to light to determine colour, the combined responses of red and green are used in a system that is sensitive purely to changes in luminance and is used to detect fine detail in a scene. For this purpose, the spectral absorbances of the two cone classes must be similar for two reasons. First, the point at which a lens focuses a light is dependent on its wavelength. If the spectral absorbances of the red and green pigments were to be moved further apart in the spectrum, the focal length for the light that primarily stimulates each of the two cone pigments would be significantly different. Therefore, when the light for one of the cone classes is in focus, the other will be out of focus (*chromatic aberration*), and if the signals from the two cone classes are

Table 3.2. *The traditional classification of colour vision abnormalities*

	Pigment basis	Effects	Genetic basis	Occurrence (%)
Dichromacies				
Protanopia	Red pigment absent	Confuses wavelengths in the range 520–700 nm	Hybrid red gene which is either inactive or produces a green pigment	M: 1.0; F: 0.02
Deuteranopia	Green pigment absent	Confuses wavelengths in the range 530–700 nm	Deleted green gene	M: 1.1; F: 0.1
Tritanopia	Blue pigment absent	Confuses wavelengths in the range 445–480 nm	Mutant blue gene	0.001–0.005; no sex difference
Anomalous trichromacies				
Protanomaly	Hybrid red pigment	Abnormal colour matches	Hybrid red gene	M: 1.0; F: 0.02
Deuteranomaly	Hybrid green pigment	Abnormal colour matches	Hybrid green gene	M: 4.9; F: 0.04

Note:
M, male; F, female.

combined, the cumulative image will be degraded. Second, if the spectral absorbances of the two cone classes differ too much, then a light of a specific wavelength and intensity may stimulate one class strongly and the other quite weakly. If one is purely looking at luminances, then one class would be signalling a strong luminance and the other would be signalling a weak luminance, which could cause problems when attempting to integrate the signals from both cone classes into a single luminance detection system.

The genetics of visual pigments

When we look at an object or scene, it is easy to assume that what we see is what everyone else sees. However, this need not be so. The different complement of visual pigments in the eyes of 'colour-blind' people means that they see a very different picture from the one most of us see. One of the first to wrestle with this problem was the chemist John Dalton. Two hundred years ago he described his own colour blindness in a lecture to the Manchester Literary and Philosophical Society, and the term Daltonism has been subsequently used to characterise this form of colour blindness. Dalton ascribed his colour blindness to a blue tint in the vitreous humour of his eye, which selectively absorbed light in the red–green range (Hunt *et al.*, 1995). A macabre twist in this tale is that he was so convinced he was right, he gave instructions that on his death his eyes should be removed and dissected to confirm his hypothesis. When he died at the ripe old age of 78 in 1844, his physician Joseph Ransome examined Dalton's eyes and found no discoloration. At this time, the main alternative to Dalton's theory was that colour blindness arose from a defect in the brain. As a result, Ransome felt bound to report that Dalton had a 'deficient development' of the phrenological organ of colour (Hunt *et al.*, 1995)!

We now know that colour blindness arises from a genetic cause and can be traced through family trees. In fact, Dalton reported that his own brother suffered the same impairment of colour vision as he did. For humans, the different forms of colour blindness are defined in terms of their difference from the 'normal' trichromacy (Table 3.2). Someone who has lost the blue pigment is called a *tritanope*, someone who has lost the green pigment is called a *deuteranope* and someone who has lost the red pigment is called a *protanope*. Those people with an altered pigment are called *anomalous* trichromats and these altered photopigments were termed anomalous pigments. For example, someone with an anomalous red pigment would be said to have protanomalous colour vision. Each normal pigment was considered to have two forms of anomalous pigment, severe and mild. The terms severe and mild reflect how much the spectral absorbance of the anomalous pigment has been altered with respect to one of the standard photopigments.

In the last few years, the techniques of molecular genetics have been used to try to determine the basis of both normal colour vision and colour blindness. Although initially these studies seemed to confirm the idea of the standard three pigment classes, more detailed studies in the last three years have shown that each of the single pigment classes is actually made up of a number of pigments with slightly different absorption spectra. Moreover, the anomalous pigments seem not to occur at one or two set positions in the spectrum but form a continuum between the spectral absorbances of the normal pigments. Therefore, what has always been called normal colour vision may just be the most commonly occurring forms of a variety of different trichromatic forms that naturally occur in the human population. In addition, the possibility must be considered that some humans may have more than three types of pigment in their eye. For example, a study published recently suggested that some women may have four cone pigments in their retina and can use them in a tetrachromatic colour vision system (Jordan & Mollon, 1993).

There is a striking sex difference in the occurrence of colour blindness: around 8% of men but only 0.5% of women seem to have abnormal colour vision. The most common causes of *inherited* colour blindness are changes to the red or green pigments; loss of the blue cone pigments is very rare. However, the most frequently occurring form of *acquired* colour blindness results from loss or damage to the blue cones, as the blue cones are extremely sensitive to high light intensities or oxygen deprivation. From the study of the inheritance of colour blindness within families, it was possible to deduce that the inheritance of the green and red pigments is *sex-linked* but that of the blue and red pigments is not (*autosomal inheritance*). However, until very recently, the nature of the genes that code for these pigments and the changes that lead to the loss or modification of the pigments could only be speculated upon. The application of the techniques of molecular genetics in the past few years is beginning to reveal the genetic basis of human colour vision and colour blindness.

All visual pigments are composed of retinal (an aldehyde derivative of vitamin A) and a protein called opsin. It is the opsin that varies in different visual pigments, and it is the opsin's structure that determines where in the spectrum the attached retinal chromophore absorbs light. In 1986, Jeremy Nathans, working at Stanford, published the first definitive study of the location and nature of the genes that code for the proteins of visual pigments (Nathans, Thomas & Hogness, 1986b). Nathans found that the gene for the blue pigment is located on *chromosome 7* and that for the red pigment is on *chromosome 3*. The green and red genes are arranged in a head-to-tail array on the *X-chromosome*. These genes show a 97% homology in their sequence, but only a 40% homology with the blue gene. The regions upstream of the green gene on the X-chromosome are also very similar to each other. As a result, it is possible for the sequence just prior to the red gene on one X-chromosome to become paired with the sequences just prior to the green pigment gene on the other X-chromosome during meiosis (see Fig. 3.3*a*) (Nathans *et al.*, 1986a). As a result, one of the X-chromosomes carries a red gene but no green gene, and the other X-chromosome carries two green genes and one red gene. A human male with the former X-chromosome lacks a green pigment and will be dichromatic. A human male with the latter X-chromosome is a 'normal' trichromat. Additional green genes could be added by similar unequal intergenic recombinations. The number of green genes on a single human X-chromosome has been reported to vary between one and six, with two being the most common complement. A research group led by Jim Bowmaker in London and John Mollon in Cambridge reported the same pattern of multiple green pigment genes in seven other Old World primate species, including the chimpanzee (Ibbotson *et al.*, 1992; Dulai *et al.*, 1994). However, no form of colour blindness has ever been observed in any Old World primate species other than humans, suggesting a strong selective pressure against a deviation from this standard trichromacy. Nathans and his colleagues suggested that the addition or deletion of red genes cannot occur in Old World primates (including humans) because the sequence following the red gene is too dissimilar to the sequence intervening between the green and red genes to pair with it. However, this finding has recently been challenged. Based on their reanalysis of published data and the use of a new technique, Jay and Maureen Neitz suggest that multiple red genes can exist on some X-chromosomes (Neitz & Neitz, 1993; 1995). If two red genes can occur on a single X-chromosome, then it follows logically that deletion or addition of red genes by unequal crossover can also occur.

Anomalous trichromats and some dichromats possess a '*hybrid*' or '*chimaeric*' gene in addition to, or instead of, one of their red or green genes (Nathans *et al.*, 1986a). This hybrid gene is composed of part of a red gene and part of a green gene. Therefore, even if Nathans is correct and the red gene cannot be deleted, dichromats lacking the red pigment may be produced when the red green gene is replaced by a hybrid gene, that is either inactive or codes for a green pigment. Hybrid genes can be formed by unequal intragenic recombination (Fig. 3.3b). The green and red genes that code for these opsins are composed of six *exons* (the sequences of DNA that code for proteins), separated by five *introns* (sequences that do not code for proteins). The differences

(a)

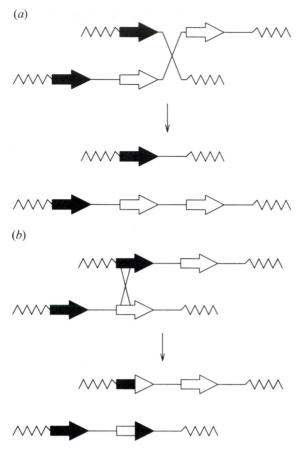

(b)

Fig. 3.3. (a) This diagram is a schematic of how intergenic recombination of the red and green genes can lead to deletion of the green gene. The red gene is shown as a black arrow and the green gene is shown as a white arrow. The wavy lines denote the rest of the X-chromosome. During meiosis, the X-chromosomes line up and exchange genetic material. Correct alignment means that they swap versions of the gene on one X-chromosome for the version on the other. At the end of the process, the chromosomes have the same numbers of genes as before. If the two X-chromosomes do not line up correctly then unequal exchange of genetic material can occur. In the case of the red and green genes, unequal crossover between misplaced red and green genes can result in the deletion of a green gene from one X-chromosome and the addition of the gene to the other X-chromosome.
(b) Sometimes, the misalignment of the X-chromosomes means that the parts of genes, and not whole genes, are exchanged (intragenic recombination). In this case, crossing over within the misplaced red and green genes can produce hybrid genes. These hybrid genes can either be inactive and so not produce a functional cone pigment, or they can produce a pigment with a spectral absorbance different from that of the red and green pigments.

between the two genes are confined to exons 2–5. So, for example, a hybrid gene composed of green exons 1–5 and red exon 6 will produce a green pigment (Merbs & Nathans, 1992a). If a male has such a hybrid gene instead of a red gene, then he will be a dichromat lacking a red pigment. If a hybrid gene has the substitution of exons 2, 3 or 4, then the spectral peak is shifted by 2–5 nm. Substitution of exon 5 shifts the peak absorbance by 15–21 nm. A trichromat with an anomalous pigment has one or more green genes and a hybrid gene with substitutions that seem to include one or more of the exons 2, 3, 4 and 5. A trichromat with an anomalous green pigment instead of the green pigment has a red gene, a hybrid gene and one or more green genes. An anomalous trichromat may then lack a green pigment but possess a green gene. The reverse situation is also found where the green gene is expressed and the hybrid gene is not. In this case the subject has normal trichromatic colour vision.

This suggests that only genes in two specific positions in the array of pigment genes on the X-chromosome will be expressed. The rest remain inactive. Further evidence for this hypothesis comes from the work of Samir Deeb and his colleagues at Washington University (Winderickx et al., 1992a). It is possible to differentiate between some green genes on the basis of a silent substitution in exon 5. A silent substitution means that there is a change in the sequence coding for an amino acid, but the new amino acid is homologous with the original and so the absorbance spectrum of the resultant photopigment is unchanged. During the process of the transcription of a protein from a gene, a mRNA copy of the gene is produced. Therefore, if a gene is active, one should be able to find these mRNA copies in a cell. If the gene is inactive, the mRNA copies will be absent. Deebs and his collegues analysed retinal cells for mRNA from subjects who have two different forms of the green gene on an X-chromosome. They found only one type of the mRNA in each subject, suggesting that only one of the two forms of the gene is expressed. The expression of only some of the photopigment genes on an X-chromosome could be the result of a genetic sequence that controls expression of the green and red genes and allows transcription of genes in only certain positions in the gene array on the X-chromosome.

Failure to express both the red and green pigments is very rare, affecting approximately 1 in 100 000. Nathans and his colleagues suggest that this deficit could arise in two ways (Nathans et al., 1989). One pathway is a two-step sequence. In the first step, unequal intergenic recombination deletes the green gene. In the second step, a point mutation inactivates the red gene. The second pathway leads to an immediate loss of green and red gene function by deletion of sequences located more distally on the q-arm than the green and red gene transcription site. These deleted sequences could either be part of an adjacent gene whose function is required for green and red cone function or they may contain elements necessary for the correct transcription of the green and red genes.

As a final codicil to Dalton's story, his eyes were preserved in a glass jar from which they must have observed the passing years with little favour. However, they have allowed us to solve a final riddle. Based on his description of his colour vision, Dalton has been classified as a protanope. However, more detailed examination of these accounts has shown that deuteranopia was an equally plausible explanation. To finally determine Dalton's colour vision, several samples of DNA were extracted from the dried-up remains of Dalton's peripheral retina (Hunt *et al.*, 1995). From these samples, it was possible to partially sequence and determine the presence of the red opsin gene. The green gene seemed to be absent, suggesting that Dalton was a deuteranope. More than 200 years after Dalton first described his condition, science has finally been able to provide an explanation of why his perception differed from normal human colour vision.

The blue cone pigment

Studies on the inheritance of deficits in the blue cone pigment in families has shown that it is inherited autosomally. Whereas red–green colour blindness is inherited *recessively*, the inheritance of the blue pigment deficit is *dominant*. Nathans' study in 1986 showed that the gene for the blue pigment is on chromosome 7, but details of the genetic deficits that caused blue blindness remained obscure. As the blue gene is not adjacent to a similar gene or stretch of genetic material on chromosome 7, it was unlikely to be lost or altered by an unequal crossover event. In 1992, Nathans and his colleagues showed that loss of the blue cone pigment can be caused by mutations at one of two sites in the blue gene (Weitz *et al.*, 1992). The mutations cause the substitution of a positively charged amino acid for a non-polar amino acid in part of the trans-membrane section of the opsin. It is believed that the effect of this non-homologous substitution is to disrupt the folding, processing or stability of the protein.

Rhodopsin, night blindness and retinitis pigmentosa

The rods mediate low intensity night vision. Other than at the very centre of the retina (the fovea) the rods dominate vision. As we have seen, there are only

6 million cones in the retina, but there are 120 million rods. Therefore, changes in rod function have a profound effect on visual perception and lead to problems such as *night blindness* and *retinitis pigmentosa* (RP). In night blindness, a person's sensitivity to light is greatly reduced, so in low light conditions that person is largely blind. RP is a far more serious condition, in which progressive degeneration of the rods and the retina eventually leads to complete blindness. It is the most common form of *retinopathy* (retinal disease) and currently afflicts up to 1.5 million people worldwide. Pathological changes associated with the disease include the progressive death of the rods and the concomitant development of night blindness (*nyctalopia*). This form of night-blindness differs from that described above, where the sensitivity of the receptors is reduced but they remain alive. In some cases, the progression of RP is rapid, and night blindness develops within the first decade of life. In other instances, such symptoms may not manifest until the fifth decade or beyond. The death of the rods is followed by more extensive pathological changes in the retina. The cone cells begin to die and there is a gradual constriction of the patient's visual fields. The retina visibly thins and the blood vessels supplying the retinal tissue begin to attenuate. Gradually, black pigmentary deposits build-up in the neural retina. This is the result of damage to the photoreceptor and pigment and epithelium layers, which leads to the migration of pigment-laden cells into the outer and inner retinal layers. Many patients eventually lose all their sight.

Inheritance of these diseases is generally autosomal dominant and is linked to a mutation of the gene that codes for the opsin component of rhodopsin. Work by Nathans' group in 1991 and Daniel Oprian's group at Brandeis University, Massachusetts in 1994 have shown that a number of mutations can occur in the rhodopsin gene that lead to changes in the opsin structure (Sung *et al.*, 1991; Rao, Cohen & Oprian, 1994). Depending upon the site of the mutation, rhodopsin can be altered in one of two possible ways. The structural changes can disrupt the folding of the protein, rendering it inactive. The change in the rhodopsin protein structure also prevents the transport and metabolic breakdown of the inactive rhodopsin; this rhodopsin accumulates in the rods leading to cell death and the retinal degeneration associated with RP. Alternatively, the mutations can prevent attachment of retinal to the mutant opsin. Under normal circumstances, when a rhodopsin molecule absorbs a photon of light it splits into retinal and an active form of the opsin, which indirectly alters the cell's membrane potential and so signals the presence of light. In the absence of light, most of the opsin molecules are bound to retinal, but there are a few unbound inactive opsin molecules. Oprian suggests that in RP, the mutant opsin cannot bind to retinal and is permanently in the active state. This has the effect of permanently stimulating the cell, which leads to cell death and ultimately to retinal degeneration. In congenital night blindness, the mutation is similar, the opsin is permanently in the active state, but it can bind retinal. When the opsin has bound retinal it does not stimulate the cell, it is only the few unbound opsin molecules that stimulate the cell. The effect of this is to partially or fully saturate the cell's response to light; unlike the situation in RP, this does not kill the cell.

Better colour vision in women?

The existence in humans of multiple pigments in the red–green region of the spectrum raises interesting questions about the colour vision of human females. With two X-chromosomes, they have two loci for each of the green and red genes. If one of the loci is occupied by a hybrid gene, then the female will have four pigments in her retina and be potentially tetrachromatic. If a green locus and a red locus are occupied by different hybrid genes then she is potentially pentachromatic. Three prerequisites must first be met before a woman with an additional pigment can use it in a functional colour vision system. First, the anomalous pigment and its 'normal' counterpart must be expressed in separate cone classes and not mixed together. Second, the nervous system must be able to discriminate between the signals from the cones with the normal pigments and those with an anomalous pigment or pigments. This may not be easy, the separation between the peak spectral absorbances of the normal pigment and that of its anomalous form can be quite small, a matter of a few nonometres. Third, the nervous system must be flexible enough to incorporate the signals from an additional cone class into a functional tetrachromatic colour system.

There is a precedence in primates for these three prerequisites. In New World monkeys, there is only a single gene locus on the X-chromosome for a visual pigment in the red–green region of the spectrum (Tovée, 1993). In a given species, there are three possible versions of the gene (*alleles*) that can occur at this locus and each codes for a cone pigment with a slightly different spectral absorbance. A male, having one X-chromosome, will have one copy of the gene and so will always be dichromatic. However, a female with two X-chromosomes can either have the same allele at both loci, and so have only a single pigment in the middle- to long-wavelength region, or have two different alleles and so have two pigments in this region. Early in embryonic development, one of the two X-chromosomes in a cell is inactivated (a process called *lyonisation*) and this inactivation is preserved in all this cell's descendants. This process is random, and so roughly 50% of the cells will have the paternal X-chromosome active and the maternal X-chromosome inactive. In the other 50% of the cells the reverse situation will be true. Lyonisation means that in a female, the products of the two X-chromosomes will be expressed in different cells. So in heterozygous females, the two different pigments will be expressed in different cells and can be used in a trichromatic visual system. A recent study at Cambridge (Tovée, Bowmaker & Mollon, 1992) has shown that in the New World monkey the common marmoset the separation between the spectral absorbances of two of the pigments in the red–green range is only 6 nm. However, in the marmosets, signals from cones containing these pigments can be distinguished and used in a colour opponent mechanism. Moreover, their nervous system is flexible enough to operate either as a di- or a trichromatic system, depending on the number of pigments present in the retina. It is, therefore, possible that the human nervous system might be sensitive and flexible enough to differentiate between, and use, four cone pigments in a tetrachromatic visual system. The additional pigment would be expressed in a separate cone class as a result of lyonisation. Experimental evidence for this hypothesis comes from further work at Cambridge by Gabi Jordan and John Mollon on a small population of putative female tetrachromats (Jordan & Mollon, 1993). Behavioural tests of their colour vision suggest that at least one of them is able to use the additional pigment in a functional tetrachromatic system.

Three pigments in normal human colour vision?

Once it was clear that hybrid genes with different proportions of the red and green genes produced pigments with different spectral absorbances, several research groups began trying to find which changes were important. They took as their starting point the hypothesis that the spectral absorbance of cone pigments in the red–green region is based on the net effect of differences in the number and position of hydroxyl groups in the vicinity of the retinal chromophore. Therefore, researchers have concentrated on changes in the gene sequence that would produce these changes. In the early 1990s, Gerry Jacobs and Jay and Maureen Neitz at Santa Barbara looked for differences in the gene sequences of cone pigments taken from humans and two New World monkey species (the squirrel monkey and the saddle-backed tamarin). They interpreted their results as suggesting that changes which caused the substitution of a single non-polar amino acid for a hydroxyl-bearing one at three sites in the opsin (positions 180, 277, 285) account for all

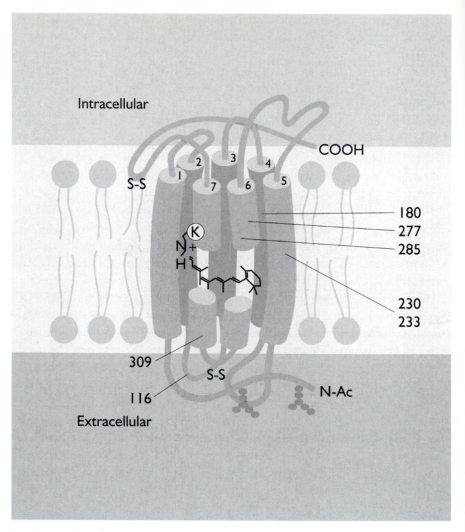

Fig. 3.4. The photopigment molecule is composed of an opsin (a protein that is located in the cell membrane) and 11-*cis* retinal (an aldehyde derivative of vitamin A). The opsins form seven α-helical, hydrophobic regions within the cell membrane, linked by straight-chain extra membrane hydrophilic loops. The membrane regions form a bundle or palisade within which retinal is bound by a Schiff's base to a lysine residue located in the centre of the seventh helix. The spectral absorbance of cone pigments in the middle- to long-wave region is thought to be based on the net effect of differences in the number and position of hydroxyl groups in the vicinity of the retinal chromophore. Five potentially important sites are indicated. Changes in the amino acids at positions 180, 277 and 285 seem to alter directly the spectral absorbance of the pigment. Substitutions at sites 233 and 230 seem to play a modulating role on the effects of changes at the other three sites.

the variation in the spectral sensitivity of cone pigments in the red–green region (Fig. 3.4) (Neitz, Neitz & Jacobs, 1991). It is suggested that the three substitutions are linearly additive in the number of nanometres by which they shift the peak sensitivity of the pigment. The substitution at 180 (located in exon 3) has been reported to shift the peak by 5.3 nm, that at 277 (in exon 5) by 9.5 nm and that at 285 (in exon 5) by 15.5 nm. However, recently, Nathans and his colleagues have been able to construct both normal and hybrid genes from complimentary DNA clones for the green and red genes and express them in cultured mammalian kidney cells (Merbs & Nathans, 1992a,b). The opsins produced by these genes are then combined with retinal to form photopigments. This technique allows the exon composition of the artificially produced genes to be controlled. For example, a gene can be produced with exon 1 from the green gene and exons 2–6 from the red gene. The production of such hybrid genes allows the effect of individual exons on the spectral absorbance of the resultant pigment to be gauged. The results show that, in addition to exons 3 and 5, altering exons 2 and 4 also changes the spectral peak of the resultant pigment. Consistent with this result is the finding in recent sequencing studies on other Old World primate species, which suggests that non-homologous substitutions at a minimum of two more amino acid sites may play a part in determining a pigment's spectral absorbance (Ibbotson et al., 1992). For example, in all the 10 primate species examined so far there is a non-hydroxyl-bearing amino acid at site 233 (exon 4) in pigments with a spectral peak above 560 nm. In the eight species with pigments having a spectral peak below 540 nm, there is a hydroxyl-bearing amino acid at 233. Moreover, the presence or absence of a hydroxyl-bearing amino acid at 233 alters the effect of non-homologous substitutions at 180. This suggests that substitutions at certain sites, such as 233, may play a modulating role on the more obvious effects of substitutions at 180, 277 and 285. To investigate this hypothesis, Oprian's group constructed a total of 28 artificial visual pigments by hybridisation and cloning techniques (Asjeno, Rim & Oprian, 1994). The red and green pigments are composed of 364 amino acid residues but differ in only 15 of these. By changing each of these amino acids in turn in the artificial visual pigments, it was possible to determine what effect an individual amino acid had on the absorbance spectra of the pigment. They found that a total of seven amino acid residues controls the spectral position of the visual pigment and not just three as suggested by Jacobs and the Neitzs (see Fig. 3.4) (Neitz et al., 1991).

A common test for red–green colour blindness is the *Rayleigh match*, in which an observer is asked to find the ratio of red to green light that matches an orange light. In the late 1980s, Jacobs and the Neitzs published results of

Rayleigh matches from populations of Caucasian males that they claimed showed a bimodality (Neitz & Jacobs, 1986). This suggested the possibility of multiple pigment types within what was conventionally classified as the red pigment. The bimodality is consistent with the results of a microspectro-photometry (MSP) study of human cone pigments by the Bowmaker and Mollon group (Dartnall *et al.*, 1983). In MSP, a beam of monochromatic light is passed through a single, isolated cone outer-segment. By varying the wavelength of the light and measuring how much is absorbed, it is possible to calculate the spectral absorbance of the cone pigment. The MSP study on the cones from seven normal trichromats showed two distinct red cone pigments. Despite this evidence, little notice was taken of this idea. The results were dismissed as experimental artefact or the result of skewed samples. However, since 1992 the idea has been vindicated. Studies of human populations show a common non-homologous substitution in the red pigment at site 180, where the amino acid can be either serine or alanine, and this difference can be correlated with the bimodal Rayleigh distribution (Merbs & Nathans, 1992b; Winderickx *et al.*, 1992b). The substitution of serine for alanine at position 180 can shift the peak absorbance of the pigment by 4–6 nm. In addition to two red pigments, it was recently reported that there may also be two green pigments that show the same substitution difference at site 180 (Neitz, Neitz & Jacobs, 1993).

Given that substitutions at several amino acid sites can alter the peak absorbance of the green and red opsins, it is possible that there are many versions of the green and red pigments naturally occurring in the human population; these may be produced by unequal intra-genic crossover or by point mutation. The so called 'anomalous' pigments would then be part of a naturally occurring range of pigments in the green and red region and what has been called normal colour vision may just be the most commonly occurring forms of a variety of different trichromatic forms that occur in the human population.

The evolution of primate colour vision

All the non-primate mammals studied so far are dichromats. They possess a blue pigment and a second cone pigment in the red–green region of the spectrum (which absorbs maximally at about 555 nm). It is believed the second gene site on the X-chromosome arose through gene duplication around 35–40 million years ago, which is roughly the time that the Old and New World primate lineages separated. The evolution of trichromacy may have arisen in one of two ways. First, a polymorphism of the pigment in the red–green region, such as now exists in the New World monkeys, may have arisen through point mutation. Then an unequal recombination event could have placed two of the different alleles on the same X-chromosome. Next, mutation may have occurred so that the two genes were expressed in different cone classes, followed by a displacement in the spectral peaks of the two cone pigments, which would produce the red and green pigments of modern Old World primates. Alternatively, the gene duplication could have involved a single form of the gene, followed by mutation and selection to produce genes at different loci coding for different pigments in different cone classes.

The colour vision system in New World monkeys may represent an intermediate step between the dichromacy of non-primate mammals and the full trichromacy of Old World primates. The cone pigment polymorphism may be maintained by the selective advantage enjoyed by the trichromatic females over the dichromatic phenotypes. However, although trichromacy is an advantage in most situations, in some situations dichromacy may be an advantage. For example, human dichromats can detect a perceptual organisation based on texture when the target is masked for normal trichromats by a rival organisation based on hue (Morgan, Adam & Mollon, 1992). The polymorphism may, therefore, be maintained by frequency-dependent advantage, such that an advantage will be enjoyed by the monkey whose vision allows detection of prey not perceived by most of his or her conspecifics.

Alternatively, the polymorphism may be a particular adaptation to the specific ecological niche occupied by New World monkeys, rather than an intermediate form on the way to the development of Old World trichromacy. Most New World monkeys live in small family groups, and there is evidence that some species, such as the saddle-backed tamarin, forage as a closely co-ordinated group. If they forage as a family group, then all the monkeys may gain if there are several forms of colour vision in the group (Tovée 1994a). Altruistic co-operation in foraging for food in family groups is favoured by kin selection. Such a situation would also select against the emergence of a uniform trichromacy. If an event occurred that expressed two non-homologous loci on one X-chromosome (and if the gene products were expressed in different cone classes), then the foraging family might be at a disadvantage because too many of its members had the same colour vision.

Key points

1. The human retina contains three classes of cone: red, green and blue. For this reason colour vision is said to be trichromatic.
2. The three cone classes are part of three opponent mechanisms. The first

takes the difference in the responses of the red and green cones (R/G), the second takes the difference between the responses of the blue cones and the sum of the red and green cones (B/Y) and the third is an achromatic mechanism that detects differences in luminance.

3. Cone pigments are composed of retinal and a protein called an opsin. The gene for the blue pigment's opsin is on chromosome 7 and those for the red and green pigments' opsins are on the X-chromosome.

4. Loss or modification of one or more of these pigments leads to an alteration in colour vision, a condition often called colour blindness. Loss or modification of the red and green cone pigments is the most common form of colour blindness, as the genes for the two pigments lie side by side and are extremely susceptible to unequal crossover during the exchange of corresponding genetic material that occurs during meiosis between a chromosome pair.

5. Unequal crossover can lead to the deletion of the red or green gene from one X-chromosome and its addition to another. A male with the former X-chromosome will be a dichromat lacking pigment encoded by the deleted gene. Unequal crossover can also cause a mixing or blending of the red and green genes to form hybrid or chimaeric genes. If these genes are non–functional, they lead to dichromacy. If they are functional, they will often produce a pigment with a spectral absorbance intermediate between that of the red and green pigments. This is called an anomalous pigment.

6. The gene for the opsin of the rod pigment is on chromosome 3, and mutations of this gene can lead to defects of rod function such as night blindness (nyctolopia) or retinitis pigmentosa.

7. The red and green pigments are composed of 364 amino acid residues but differ in only 15 of these residues. A total of seven of these residues control the peak spectral position of the visual pigment in the red–green range.

8. There may be many versions of the green and red pigments naturally occurring in the human population, and the so called 'anomalous' pigments would then be part of a naturally occurring range of pigments in the green and red region. What has been called normal colour vision may just be the most commonly occurring form.

4

The organisation of the visual system

Making a complex process seem simple

Vision is the primary sensory modality in primates such as ourselves, and this is reflected in the complexity of the visual system and the extent of the cerebral cortex used for the analysis of visual information. On the basis of anatomical, physiological and behavioural studies, it is believed that at least 32 separate cortical areas are involved with the processing of visual processing in the macaque monkey (Van Essen, Anderson & Felleman, 1992). Of these areas, 25 are primarily visual in function; the remaining seven are also implicated in other functions such as polysensory integration or visually guided motor control. These visual areas occupy about half of the $100 \, cm^2$ area of each of the monkey's cortical hemispheres. Two of the areas, V1 and V2, each exceed more than $10 \, cm^2$ of the cortical surface, but most visual areas occupy less than a tenth of this size. Comparatively little is known of the functional anatomy of the human visual cortex, but it seems to be at least as complex as that of the monkey (Kaas, 1992; Sereno et al., 1995). Fortunately, it is possible to simplify this picture by concentrating on the key visual areas and looking at their functional organisation.

As one moves up the visual system, from the retina to the lateral geniculate nucleus (LGN) and then on to successive cortical areas, visual neurons become responsive to more and more complex stimuli. For example, in monkeys, in the first cortical visual area (called primary visual cortex or V1) there are neurons responsive to simple lines of different orientations, whereas in one of the higher visual areas (inferior temporal cortex) the neurons respond to complex stimuli, such as faces. However, the visual system is not organised in just a serial, hierarchical pathway. Different aspects of a stimulus (such as its shape, colour and motion) are analysed in separate, parallel pathways. These pathways are usually divided into two broad categories; *'what'* and *'where'* pathways. The 'what' pathway deals with information about the stimulus features (such as shape and colour) and the identity of an object; it can be subdivided into two further pathways: colour and shape. The 'where' pathway deals with spatial information about an object and is usually subdivided into motion and form derived from motion. This chapter will first describe the basic anatomy of the visual system and its connections, and then it will explain how this machinery functions to produce vision.

The retina

As stated above, the visual system can be divided into two or more separate pathways. This separation starts to become evident at the level of the retina. There are several forms of ganglion cell in the primate retina, of which two types (M and P cells) constitute about 90% of the cells. The *M class* (sometimes also called A or Pα cells) account for 10%, and the other 80% are accounted for by the *P class* (sometimes called B or Pβ cells). The subsequent pathways are often referred to as the *M or P pathways*. The remaining 10% of the retinal ganglion cells consist of at least eight different types (Rodieck, 1988). The P cells are selective for wavelength and high spatial frequencies and have slow sustained (*tonic*) responses, whereas the M cells are not wavelength sensitive but are sensitive to low spatial frequencies, have transient (*phasic*) responses and have faster conduction velocities. At any one eccentricity, the dendritic field of the M cell is three times larger than that of the P cell. These differences in the response properties of the neurons shape the functions of the subsequent visual areas.

The lateral geniculate nucleus

The axons of all of the ganglion cells come together to form the *optic nerve*, which passes out of the eye through the optic disc and projects to the *dorsal LGN* (Fig. 4.1). The nerves from the two eyes join together before they reach the LGN, to form the *optic chiasm*. At this point, axons from ganglion cells from the inner parts of the retina (the nasal sides) crossover and then continue on to the LGN. As the axons from the nasal portions of the retinae cross

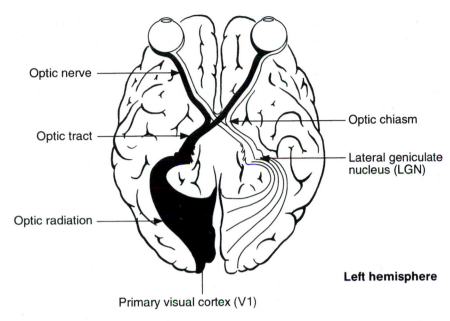

Optic nerve

Optic chiasm

Optic tract

Lateral geniculate
nucleus (LGN)

Optic radiation

Left hemisphere

Primary visual cortex (V1)

Fig. 4.1.   The connections from the retina to the cerebral hemispheres. (Modified and redrawn from Zeki, 1993.)

to the other side of the brain, each hemisphere of the brain receives information from the opposite side of the visual scene. So if you look straight ahead, the right hemisphere receives information from the left half of the visual field, and the left hemisphere receives information from the right side of the visual field.

The LGN is a folded sheet of neurons, about the size of a credit card but about three times as thick, found on each side of the brain (Fig. 4.2). It consists of six layers. Each layer receives input from only one eye, layers 2, 3 and 5 from the eye on the same side as the LGN, and layers 1, 4 and 6 from the eye on the opposite side. The topographic arrangement of the ganglion receptive fields is maintained in the LGN, so that each layer contains a complete map of the retina. The cells in layers 1 and 2 contain larger cell bodies than those in the remaining four layers. The inner two layers are, therefore, called *magnocellular* or M layers and the outer four layers are called *parvocellular* or P layers. Neurons in the M layers receive input from M ganglion cells and neurons in the P layers receive input from type P ganglion cells. The LGN neurons show the same response characteristics as the cells from which they receive input.

LGN

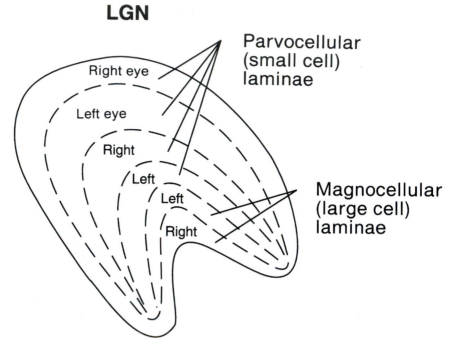

Fig. 4.2. Schematic diagram of a coronal section through the LGN from the left hemisphere of the macaque monkey, showing the six layers. Magnocellular layers 1 and 2 receive input from retinal M cells, while parvocellular layers 3 through 6 receive input from P cells. Cell bodies in layers 1 and 2 are larger than those found in layers 3 through 6, hence the names of the two types of layers. (Redrawn from Lennie & D' Zmura, 1988.)

The primary visual cortex

The LGN neurons mainly project to the *primary visual cortex* (also known as the striate cortex or V1). This is the first cortical visual area and consists of six principal layers (and several sub layers) arranged in bands parallel to the surface of the cortex. The axons from the LGN terminate on cortical neurons in layer 4 (Fig. 4.3). The P layer neurons send their axons to neurons in the deeper part of this layer (sub layer 4Cβ), which in turn send their axons to layers 2 and 3 and from there to V2. The M layer neurons send their axons to neurons in sublayer 4Cα, and the information is then relayed to layer 4B and then to V2 and to V5 (Fig. 4.4). Cells in layer 4B are orientation selective and most show selectivity for the direction of movement. Some of these neurons

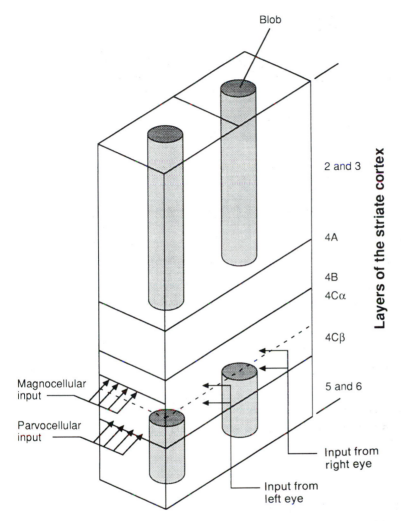

Fig. 4.3. A schematic diagram of a small part of V1, showing two blobs and their inter blob region.

are binocular (require stimulation from both eyes) and show sensitivity to retinal disparity (the difference in the relative position of the stimuli in the visual field of two eyes) (Poggio & Fisher, 1977).

The P pathway splits to produce two new pathways in the upper layers of V1. One pathway seems to deal primarily with colour and this is called the *P–B pathway*. Neurons in the second pathway are sensitive to features such as the orientation of the stimulus and seem to mediate high acuity perception. This pathway is called the *P–I pathway*. This separation been demonstrated

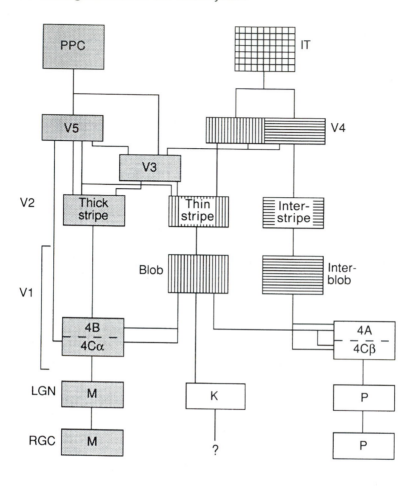

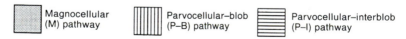

Magnocellular (M) pathway

Parvocellular–blob (P–B) pathway

Parvocellular–interblob (P–I) pathway

Fig. 4.4. Subcortical and cortical pathways in the macaque. There are two main pathways: the parvocellular (P) pathway and magnocellular (M) pathway. The P pathway splits to produce two new pathways in the upper layers of V1. One pathway seems to deal primarily with colour and this is called the P–B pathway. Neurons in the second pathway are sensitive to features such as the orientation of the stimulus and seem to mediate high acuity perception. This pathway is called the P–I pathway. The M pathway is dominated by a single source, but the P–I and P–B streams receive inputs from a number of different sources. At the top of the cortical hierarchy, the M pathway leads primarily to the posterior parietal cortex (PPC), which processes spatial and motion information. The P–I and P–B pathways project to the inferior temporal cortex (IT), which mediates pattern and object recognition. RGC, retinal ganglion cell. (Redrawn from Van Essen & Deyoe, 1995.)

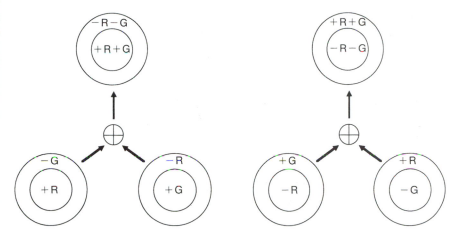

Fig. 4.5. A schematic illustration of how a cell with double colour-opponent properties could be derived from colour opponent cells. (Redrawn from Lennie & D'Zmura, 1988.)

anatomically by staining for the mitochondrial enzyme cytochrome oxidase. In both monkeys and humans, this stain shows an arrangement of dark-staining columns that extend through layers 2 and 3 and, more faintly, through layers 5 and 6 (Wong-Riley, 1993). These columns, called *blobs*, are 0.2 mm in diameter and spaced at roughly 0.5 mm intervals (Fig. 4.3). The areas surrounding the blobs are called the *interblob region*. In the interblob region, most cells respond to stimuli of a particular orientation, such as lines or bars, and have small receptive fields. Most of these cells do not show colour-coded responses. They show no colour opponency and respond well to achromatic luminance-contrast borders. This suggests that the colour-coded P cell input is pooled in such a way that colour contrast can be used to identify borders, but that the information about the colours forming the border is lost. The neurons in the interblob region are part of the P–I pathway.

The blob cells are not orientation selective but are either colour or brightness selective. These cells are part of the P–B pathway. The P–B pathway, therefore seems to carry information complementary to the information carried by the P–I pathway. The colour–opponent blob cells receive input from the colour-opponent P cells in the LGN, although they differ in that their receptive field centres are larger and their colour coding is double-opponent. (They give opposite responses to different parts of the spectrum in the different parts of their receptive field. For example, the centre might give an ON-response to green and an OFF-response to red, and the opposite set of responses in the surround.) A possible means of deriving double-opponent cells from opponent cells is shown in Fig. 4.5.

The blob and interblob systems, therefore, seem to work in different but complementary ways. The blob cells are colour coded, excited by colours in one region of the spectrum and inhibited by others, and are not selective for stimulus orientation. Interblob cells are selective for stimulus orientation but mostly are not colour selective, responding to a line or edge of the correct orientation regardless of its colour.

Visual area 2

The main target of V1 is V2. Staining for cytochrome oxidase in this area does not reveal a pattern of blobs or interblobs, but instead a pattern of stripes running perpendicular to the border between V1 and V2 and extending over the entire 8 to 10 mm width of V2. There seems to be three types of stripe. There are two darkly staining stripes, one *thick* and one *thin*, separated by more lightly staining *interstripes* (sometimes called *pale stripes*). The neurons in layer 4B of V1 (part of the M pathway) project to the thick stripes (Fig. 4.4). Neurons in the thick stripes show similar response properties to the neurons in layer 4B. They are orientation and movement selective and many show sensitivity to retinal disparity (the basis of stereopsis). The neurons in the blobs project to the thin stripes; neurons in the thin stripes are not orientation selective, but more than half are colour sensitive (mostly double-opponent). The interblobs project to the interstripes; neurons in this region are orientation selective but not direction selective, nor do they show colour selectivity. The organisation of V1 is retinotopic, that is the visual field of the retina is mapped onto the surface of the cortex of V1. In V2 there seems to be three separate visual maps (Roe & Ts'o, 1995). Within the thick stripes, there is a visual orientation map; within the thin stripes there is a colour map; and within interstripes a disparity map. Adjacent stripes are responsive to the same region of the visual field. So there are three, interleaved visual maps in V2, each representing a different aspect of the visual stimulus.

The M pathway projects from layer 4B of V1 to the thick stripes of V2. The P–B pathway projects from the blobs of V1 to the thin stripes of V2 and the P–I pathway projects from the interblob region to the interstripes.

Visual area 4

Both subdivisions of the P pathway, the thick stripes (colour) and interstripes (form) project to V4. V4 and the other visual areas upstream of V2 all stain relatively homogeneously for the cytochrome oxidase enzyme, and no alternative marker is yet known. However, the continued separation of the two subdivisions of the P pathway in V4 can be inferred from patterns of connectivity. It is possible to trace connections by using substances, such as peroxidase enzymes or dyes, that will be absorbed by the neurons and transported up or down their axons. This method reveals the neuronal connections of the specific piece of cortex in which the tracer is deposited. When tracer is deposited in different parts of V4, the back-filled neurons in V2 tend to occur in one of two distinct patterns, either they are largely restricted to the thin stripes or they are largely restricted to the interstripes (Shipp & Zeki, 1985; Zeki & Shipp, 1989). This suggests that the separation of the two subdivisions of the P pathway continues in V4.

Single-cell recording studies suggest that some of those V4 cells that are sensitive to colour seem to display a higher sophistication of response than those neurons in preceding areas. The neurons respond not to the wavelength of light but to its 'colour' (Zeki, 1983), a phenomenon known as colour constancy (see Chapter 7). Lesions of V4 seem to impair the ability of monkeys to identify colours when illumination conditions are changed despite the fact that they are able to detect very small differences in wavelength (Wild *et al.*, 1985). Damage to the equivalent area in humans also impairs the ability to distinguish colour, a condition called *achromatopsia*. Interestingly, damage to certain parts of human V1 and V2 causes the opposite condition (*chromatopsia*), where people can see colours but not shape or form. V4 is also important for object discrimination. Single-cell recording studies in monkeys suggest that some V4 cells are sensitive to simple shapes and objects, and V4 lesions impair object discrimination. V4 projects primarily to the *temporal visual cortex*, where there seems to be an integration of form and colour to give a representation of complex objects. Neurons in this region are responsive to complex patterns and objects, such as faces (Rolls & Tovée, 1995).

Visual areas 3 and 5

The M pathway projects to V3 and V5, both directly from layer 4B of V1 and through the thick stripes of V2. Most cells in V3 are orientation selective and are believed to be concerned with processing dynamic form. V5 (also known as the middle temporal visual area or MT) is believed to process information on motion and stereoscopic depth. In monkeys, lesions of V5 cause deficits in pursuit eye movements and in discriminating the direction of motion. The M pathway then projects to the *parietal cortex*. The parietal cortex seems to be important for the integration of movement and depth into a representation of space. Damage to this region in humans causes a condition called Balint's syndrome. This has three main symptoms. First, a difficulty in reaching for objects under visual guidance (*optic ataxia*). Second, subjects display a deficit in visual scanning (*ocular apraxia*). A person may perceive an object normally but will be unable to maintain fixation. His or her eyes will wander. He or she will not be able to make a systematic scan of a scene and will not be able to perceive the location of the objects seen. Finally, the subject is only able to see one object at a time in a crowded scene (*simultagnosia*).

The separation of the pathways

The separation of the P and M pathways should not be overemphasised. There is some communication between the two pathways. If the M layers of the LGN are inactivated by cooling, the visual responses of neurons are reduced both in V4 and in the blobs and interblobs of V1 (Nealey & Maunsell, 1994; Ferrera, Neally & Maunsell, 1994). This suggests that there is an input from the M pathway into the P pathway. Moreover, there is a third type of cell in the LGN (Fig. 4.4). These small neurons are called *K* or *W cells* and are found within the gaps between the M and P layers. They are believed to receive input from $P\gamma$ cells, one of the cell types that make up the remaining 10% of retinal ganglion cells. The W neurons also project to the blobs of V1 (Hendry & Yoshioka, 1994). Therefore, the so-called P pathway actually receives input from the P, M and W neurons of the LGN. The M pathway does seem to be more segregated. For example, inactivation of the P layers of the LGN has a negligible effect on the responses of neurons in V5 (Maunsell, Nealey & DePriest, 1990).

The functional organisation

It has been proposed that, in Old World primates, visual information is processed in two broad systems: the 'what' system (also called the *ventral* system), which is concerned with the identification of an object, and the 'where' system (also called the *dorsal* system), which is concerned with the relative spatial position of an object (Mishkin, Ungerleider & Macko, 1983) (Fig. 4.6). Damage to the 'what' system, through which V1 projects to the temporal lobes, does not impair performance of visuospatial tasks but does impair performance of object discrimination tasks. Damage to the 'where' system,

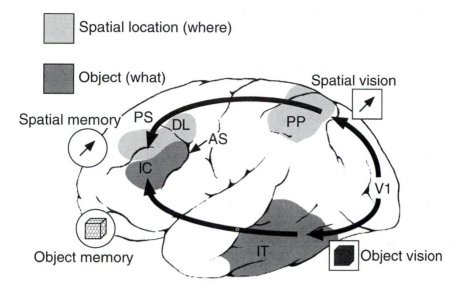

Fig. 4.6. Schematic diagram illustrating the location of the 'what' and 'where' pathways in the primate brain. PS, principal sulcus; AS, arcuate sulcus; PP, posterior parietal cortex; IT, inferior temporal cortex; DL, dorsolateral frontal cortex; IC, inferior convexity of the frontal cortex. (Redrawn from Wilson *et al.*, 1993.)

through which the V1 projects to the parietal lobes, produces impairments of visuospatial tasks but does not impair the performance of object discrimination tasks. The visual cortex of New World monkeys appears to be similarly organised into separate 'what' and 'where' streams (Weller, 1988). This suggests a common primate plan, which PET scan studies suggest extends to humans (Haxby *et al.*, 1991).

The two streams project to different prefrontal cortical areas (Wilson, O'Scalaidhe & Goldman-Rakic, 1993). The 'what' system projects to the cortex of the *inferior convexity* (IC) ventrolateral to the principal sulcus and the 'where' system projects to the *dorsolateral* prefrontal region (DL). The prefrontal cortex is an important region for working memory. Patricia Goldman-Rakic and her colleagues trained monkeys to carry out visual tasks, while at the same time they recorded from single neurons in either the IC or DL. In the first task, the monkeys were trained to stare at a spot on a video screen while an image flashed at one of several locations on the screen and then disappeared. A few seconds later, a cue on the screen signalled the monkeys to move their gaze to where the image had been, indicating that they had remember the location of the image. In the second task, the location of the image remained constant, but the image itself changed. The monkeys were trained to wait until the image disappeared and then, after a delay, move their eyes to the right if they saw one image and to the left if they saw another, indicating they remember information about an object's features. In the first task, neurons in DL became active during the delay period, while neurons in IC did not alter their activity. However, during the second task the pattern of activity is reversed; the neurons in IC are active during the delay and the neurons in DL remain quiet. These results suggest that IC mediates working memory for objects and Dl mediates spatial working memory.

The M and P pathways have been thought to correspond roughly to these two systems (Livingstone & Hubel, 1988). The P pathway, with information about colour and shape, would seem to be ideal for the 'what' system, which develops a representation of an object. Similarly, the M pathway, with information about motion, stereopsis and form derived from motion, would seem to be the obvious candidate for the 'where' system, which develops a representation of spatial relationships in the visual field. However, there is considerable communication between the two systems at all levels (Harries & Perrett, 1991).

Perception versus action

An alternative approach to the 'what' versus 'where' organisation has been proposed by Goodale & Milner (1992), which takes more account of the output requirements of the system. This approach is called 'what' versus 'how'. Visually guided grasping was studied in a patient with Balint's syndrome (as mentioned above, in this syndrome bilateral parietal damage causes profound disorders of spatial attention, gaze and visually guided reaching). While this patient had no difficulty in recognising line drawings of common objects, her ability to pick up objects remained impaired (Jakobson et al., 1991). Not only did she fail to show normal scaling of the grasping movement, she also made a large number of adjustments in her grasp as she closed in on an object. These adjustments are rarely observed in normal subjects. Such studies suggest that damage to the parietal lobe can impair the ability of patients to use information about the size, shape and orientation of an object to control the hand and fingers during a grasping movement. Another patient developed profound *visual-form agnosia* (an inability to recognise objects) following carbon monoxide poisoning (Goodale et al., 1991). Despite her profound inability to recognise the shape, size and orientation of an object, the patient showed strikingly accurate guidance of hand and finger movements directed at the very same objects. So despite impaired conscious visual discrimination of the objects, visual information, computed unconsciously, was made available to the action system to direct grasping actions.

Parietal damage can also result in an impairment of the ability to see more than one object in one's visual field (simultagnosia), suggesting an important role for the parietal cortex in spatial attention. However, unconscious perception may also occur in this condition too. Castiello and his colleagues report a simultagnosic patient who was able to respond to the presence of more than two objects in her environment, but only when the two objects were semantically related to each other; for example, if the two objects might normally be used together or if the two objects were from the same category (Castiello, Scarpa & Bennett, 1995). When the objects were unrelated, the patient was unable to respond to them. Despite the effects of the related objects on her actions, the patient denied seeing both of the objects. These results suggest an unconscious perception of both objects and relations between objects, which can affect actions.

Goodale and Milner's theory (1992)has been criticised for the small number of patients involved in the studies on which it is based and for the failure to determine accurately the brain areas mediating their patient's surviving abilities (Ungerleider & Haxby, 1994). It has been argued that Goodale and Milner's results fail to address the question of whether patients with parietal damage demonstrate visuospatial impairments that cannot be attributed to a primary visuomotor defect. Von Cramon & Kerkhof (1993) have reported that 67 patients with focal parietal lesions (confirmed by MRI), had impaired perception of horizontal and vertical axes, poor length and distance estimation and a deficit in orientation discrimination, as well as a deficit in position matching. There is also evidence that impaired perception of axes and angles was associated with anterior parietal damage, whereas impaired perception of position and distance was more closely associated with posterior parietal damage. This evidence would suggest that although visuospatial deficits may be accompanied by visuomotor impairments they are not explained by them (Ungerleider & Haxby, 1994).

Blindsight

Damage to V1 causes holes (*scotomas*) in our visual field. People with V1 damage seem to have no conscious perception of the visual stimuli presented in these scotomas. As we have seen in the preceding sections, all the main visual pathways pass through V1, and so damage to this bottleneck might be expected seriously to disrupt vision. However, some patients can respond to visual stimuli presented in their scotomas if they are required to make a forced choice to indicate stimulus parameters. Some patients are able to look towards stimuli presented in their scotomas, to localise them by pointing and to detect and discriminate movement (Cowey & Stoerig, 1991). One patient with complete cortical blindness was able to follow a large moving striped display with his eyes, despite disclaiming any visual sensation that might explain his visual tracking. Patients can detect and discriminate flicker, orientation and wavelength. Their pupils continue to respond to changes in light level, pattern and contrast, and when asked to reach for visual targets, two patients adjusted their grasp so that it matched the size and shape of the object. They could also use the meaning of words flashed in their blindfields in order to select between pairs of words subsequently presented in their intact field. This perception of visual stimuli without conscious knowledge is called *blindsight*.

The primary projection from the retina is to the LGN, and this exceeds a million fibres per eye. However, the retina also projects to other structures, such as the projection to the *superior colliculus* (SC) (approximately 100 000–150 000 fibres) (see Fig. 4.7). Many of these connections transmit information about the position, size and movement of visual stimuli (Cowey & Stoerig, 1991). It is believed these connections may mediate the residual vision found in blindsight. This hypothesis is supported by the effects of lesions of these subcortical pathways in monkeys. Monkeys also seem to show blindsight (Cowey & Stoerig, 1995) and so provide an animal model for investigating this phenomenon. Damage to V1 has profound effects on the subsequent visual areas. Lesions of either the SC or the *lateral pretectum* (with interruption of the accessory optic system) drastically reduce the blindsight capacities of monkeys who had V1 removed (Mohler & Wurtz, 1977; Pasik & Pasik, 1982).

Doubts have been raised as to whether blindsight does actually exist. Alternative suggestions have included the possibility that the patients' V1 was merely damaged not destroyed, that patients responded to light scattered from the stimulus onto the intact retina, or that experimenters employed a lax criterion for detection that was very different from the one used in the normal field. For at least some of the studies' these reservations are groundless. First blindsight can be demonstrated even in patients in whom V1 or even the entire cerebral hemisphere has been surgically removed. Second, when a stimulus that is detected in blindsight is presented on the natural blindspot (the optic disk), it becomes undetectable, despite the fact that the optic disc normally scatters more light than the rest of the retina. Third, a stimulus presented in the blind field and to which the patient is not even asked to respond can influence the response to a companion stimulus presented in the intact field. It is true that in the case of one patient who was diagnosed as having blindsight an MRI scan indicated the presence of some remnants of V1 that might have

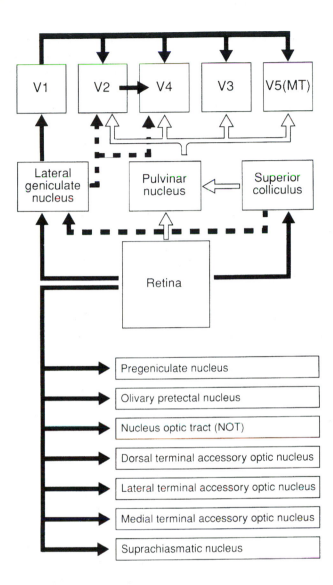

Fig. 4.7. Fibre pathways from the retina of the eye to the brain of primates. The primary pathway is through the LGN to V1. The dashed lines indicate particularly sparse pathways. None of the seven pathways shown at the bottom have direct connections to any of the cortical areas shown at the top, although indirect connections may exist. Visual areas V1, V2, V3, V4 and V5 all have additional connections not shown, with each other and with further visual areas. (Redrawn from Cowey & Stoerig, 1993.)

mediated some of the residual vision (Fendrich, Wessinger & Gazzaniga, 1992); however, there is also considerable evidence of patients studied at autopsy, during surgery or by non-invasive scanning who had complete loss of V1 (see Cowey & Stoerig, 1993).

A slightly different criticism of the blindsight concept has come from recent work which suggests that there may be conscious perception of movement in human subjects with damage to, or reversible inactivation of, V1 (Barbur et al., 1993; Beckers & Zeki, 1995). It has been suggested that there may be a 'fast' pathway that bypasses V1 and connects V5 directly with either the LGN or possibly even the retina. (Beckers & Zeki, 1995). The information conveyed by this fast pathway seems to be sufficient to mediate a conscious, coarse perception of movement and would enable an individual to react more quickly to moving stimuli than if the information came via V1. This hypothesis is supported by a visual evoked potential study which suggests that to some forms of moving stimuli, V5 may become active before, or simultaneously with, V1 (ffytche, Guy & Zeki, 1995). This is not to suggest that V1 is not an extremely important stage in the processing of movement information. The fast connection would represent only a small percentage of V5's connections, and input from connections that come through V1 are required for any form of detailed movement perception and discrimination.

The fast connection hypothesis has limited support in monkey studies. Neurons in V5 continued to respond when transmission through V1 is blocked (Rodman, Gross & Albright, 1989a; Girad, Salin & Bullier, 1992), but ablation of the SC largely abolished this activity (Rodman, Gross & Albright, 1989b). This connection is probably related to the need to integrate eye movement (in which the SC plays a role) with external motion as processed by V5. Therefore, although there is evidence of connections that bypass V1 on their way to V5, there is little evidence for the suggested fast pathway. The role of V1 seems much more important in the activity of other extrastriate visual areas. Girard and Bullier (1988) cooled V1 and tested visual responsiveness in V2, where they found less than 2% of neurons responded normally. This is consistent with a similar study in V4 (Schiller & Malpeli, 1978), suggesting these areas are extremely dependent on V1 for visual information.

Key points

1. The visual system is divided into two or more streams of information, in which different aspects of visual perceptions such as movement, depth, colour and shape are processed separately.

2. This division is first evident at the level of the retinal ganglion cells. There are two main classes of ganglion cell: the M class (which gives rise to the M pathway) and the P class (which gives rise to the P pathway).

3. The axons of the ganglion cells form the optic nerve, which projects to the dorsal lateral geniculate nucleus (LGN). The nerves from the two eyes join before they reach the LGN to form the optic chiasm. At this point, axons from ganglion cells from the inner portions of the retina (the nasal sides) crossover and then continue on to the LGN, so each hemisphere of the brain receives information from the opposite side of the visual scene.

4. The LGN consists of six layers. Each layer receives input from only one eye: layers 2, 3 and 5 from the eye on the same side as the LGN, and layers 1, 4 and 6 from the eye on the opposite side. The cells in layers 1 and 2 contain larger cell bodies than those in the remaining four layers. So cells in layer 1 and 2 are called magnocellular (M) and those in the other layers are called parvocellular (P). Neurons in the M layers receive input from M ganglion cells and neurons in the P layers receive input from type P ganglion cells.

5. The LGN neurons mainly project to the primary visual cortex (V1). This is the first cortical visual area and consists of six principal layers (and several sub layers) arranged in bands parallel to the surface of the cortex. The axons from the LGN terminate on cortical neurons in layer 4.

6. The P layer neurons send their axons to neurons in the deeper part of this layer (sub layer 4Cβ), which in turn send their axons to layers 2 and 3 and from there to V2.

7. The M layer neurons send their axons to neurons in sublayer 4Cα, and the information is then relayed to layer 4B and then to V2 and to V5. Cells in layer 4B are orientation selective and most show selectivity for the direction of movement. Some of these neurons are binocular and show sensitivity to retinal disparity.

8. The P pathway splits to produce two new pathways in the upper layers of V1. One pathway seems to deal primarily with colour and this is called the P–B pathway. Neurons in this pathway are found in columns called blobs. Neurons in the second pathway are sensitive to features such as the orientation of the stimulus and seem to mediate high acuity perception. This pathway is called the P–I pathway. These neurons are found in the area surrounding the blobs (the inter blob region).

9. The main target of V1 is V2. The neurons in layer 4B of V1 (part of the M pathway) project to the thick stripes, and neurons in this stripe are orientation and movement sensitive. The neurons in the blobs project to

the thin stripes; neurons in the thin stripes are not orientation selective but more than half are colour sensitive (mostly double opponent). The interblobs project to the interstripes; neurons in this region are orientation selective, but not direction selective or colour selective.

10. Both subdivisions of the P pathway, the thick stripes (colour) and interstripes (form), project to V4, and the two streams remain separate in this area. Some V4 cells respond not to the wavelength of light but to its 'colour', a phenomenon known as colour constancy. Damage to the equivalent area in humans also impairs the ability to distinguish colour, a condition called achromatopsia. V4 is also important for object discrimination. V4 projects primarily to the temporal visual cortex, where there seems to be an integration of form and colour to give a representation of complex objects.

11. The M pathway projects to V3 and V5, both directly from layer 4B of V1 and through the thick stripes of V2. Cells in V3 are orientation selective and are believed to be concerned with processing dynamic form. V5 (or MT) is believed to process information on motion and stereoscopic depth. The M pathway then projects to the parietal cortex, which is important for the integration of movement and depth into a representation of space. Damage to this region in humans causes a condition called Balint's syndrome. This has three main symptoms; optic ataxia, ocular apraxia and simultagnosia.

12. Visual information is processed in two broad systems: the 'what' system (also called the ventral system), which is concerned with the identification of an object, and the 'where' system (also called the dorsal system), which is concerned with the relative spatial position of an object. Another approach is to divide the pathways into 'what' versus 'how'. In this scheme, the parietal cortex is primarily concerned with visual cues important for the conversion of visual cues into spatial information for motor movement and interaction with the environment.

13. Damage to V1 causes blindness, but under forced-choice conditions a patient may display perception of visual stimuli without conscious knowledge. This is called blindsight. It is believed that this perception is mediated by subcortical pathways that bypass V1 and project directly to the later visual areas.

5

The primary visual cortex

The visual equivalent of a sorting office?

The primary visual cortex (V1) or striate cortex is an important area in which partially processed information from the retina and LGN is separated and packaged up for more elaborate analysis in the specialised visual areas of the extrastriate cortex. But V1 is more than just a neural version of a post office sorting department. The response properties of most neurons in V1 are very different from those of neurons in the preceding area. New response features, such as sensitivity to lines and bars of different orientations and movements, are created, along with a specialisation of some neurons to an existing visual feature such as colour. Moreover, the functional organisation of V1 into repeating columns and modules seems to be a standard pattern in all cortical visual areas; this pattern of organisation is an efficient way of mapping a multi-dimensional stimulus, such as vision, onto an irregularly shaped piece of two-dimensional cortex.

Visual information passes to the cortex from the LGN through the optic radiation. In the monkey, the first cortical visual area (V1) consists of a folded plate of cells about 2 mm thick, with a surface area of a few square centimetres. This is a much larger and more complex structure than the LGN; for example, the LGN is composed of 1.5 million cells, whereas V1 is composed of around 200 million. V1 lies posteriorly in the occipital lobe and can be recognised by its characteristic appearance. Incoming bundles of fibres form clear stripes in this area, hence the name striate cortex. Adjacent regions of cortex are also concerned with vision. The area that immediately surrounds V1 is called V2 (sometimes called area 18) and receives input primarily from

V1. Each area contains its own representation of the visual field projected in an orderly manner. The topography of the visual field is preserved in the projection to V1 from the LGN, with relatively more cortex being devoted to the fovea (about $300\,mm^2/degree^2$) than to the peripheral visual field (about $0.1\,mm^2/degree^2$ at 20 degrees of retinal eccentricity).

Segregation of layer 4 inputs

A general feature of the mammalian cortex is that the cells are arranged in six layers within the grey matter. These layers vary in appearance from area to area in the cortex depending on the density of cell packing and the thickness. In V1, most of the incoming fibres from the LGN terminate in layer 4 (see Fig. 4.3 p. 63). This layer is subdivided into three sublayers, A, B and C. Sublayer C is further subdivided into $4C\alpha$ and $4C\beta$. Projections from the LGN's parvocellular layers terminate in sublayers 4A and $4C\beta$ and the upper part of layer 6. The cells in layer $4C\beta$ then supply cells in layers 2 and 3. The magnocellular layers terminate in layer $4C\alpha$ and the lower part of layer 6. The cells in layer $4C\alpha$ supply layer 4B. Both the P and M pathways send inputs to the blobs. Thus, the separation into P and M pathways first observed in the retina is preserved in V1.

Also preserved in layer 4 is the separation of inputs from the two eyes. In cats and monkeys, if the cells in one layer of the LGN receive their input from one eye, the next layer will receive inputs from the other eye. The cells from one LGN layer will project to groups of target cells in layer 4C, separate from those supplied by the other eye. These groups of cells form alternating stripes or bands in layer 4C. Above and below this layer, most cells are driven by both eyes, although one eye is usually dominant. Hubel & Wiesel (1977) termed these blocks of cells *ocular dominance columns*.

Evidence for this alternating projection comes from a number of sources. First, if a small lesion is made in one layer of the LGN, degenerating terminals subsequently appear in layer 4 in a characteristic pattern of alternating stripes (Hubel & Wiesel, 1977). These correspond to areas driven by the eye in whose line of connection the lesion is made. Further, if one injects a radioactively labelled amino acid, such as proline or leucine, into the vitreous humour of one eye, it is taken up by the nerve cell bodies of the retina and incorporated into a protein. The labelled protein is then transported from the ganglion cells through their projections to the LGN, and then on through the projections of LGN cells to layer 4C. The striped arrangement demonstrated in this way is the same as that produced by lesions of the LGN, but now the pattern is in all parts of the primary visual cortex. The colour-selective blobs stained by cytochrome oxidase (described in Chapter 4) lie in the centre of each ocular dominance column.

Cortical receptive fields

There seem to be two broad categories of neurons in V1: termed *simple* and *complex* neurons. In addition, there is a class of cell found exclusively in layer 4C (where most LGN fibres terminate), that has a concentric centre–surround receptive field like that of the LGN and the retinal ganglion cells.

Simple cells are found mostly in layers 4 and 6, and these two regions receive direct inputs from the LGN. One type of simple cell has a receptive field that consists of an extended narrow central portion flanked by two antagonistic areas. The centre may be either excitatory or inhibitory. For such cells, optimal activation is by a bar of light that is not more than a certain width, that entirely fills the central area and that is oriented at a certain angle. The optimal width of the narrow light or dark bar is comparable to the diameters of the ON-centre or OFF-centre regions of the centre–surround receptive fields of the LGN and ganglion cells (Fig. 5.1). Resolution has not been lost but has been incorporated into a more complex receptive field. The preferred orientation of the light or dark bar varies with different cells, as does the symmetry of the receptive field. In some cases, the receptive field may just consist of two longitudinal regions facing each other, one excitatory and one inhibitory. Another type of cell is sensitive to the length of the stimulating bar. There seems to be an additional antagonistic area at the top or bottom of the receptive field. As a result, the optimal stimulus is an appropriately oriented bar or edge that stops in a particular place. These cells are *end-inhibited* or *end-stopped* cells. In spite of the different proportions of inhibitory and excitatory areas, the two contributions match exactly and so diffuse illumination of the entire receptive field produces little or no effect.

Complex cells are abundant in layers 2, 3 and 5. In common with simple cells, these cells require a specific orientation of a dark–light boundary, and illumination of the whole receptive field is ineffective (Fig. 5.2). However, the position of the stimulus within the receptive field is not important as there are no longer distinct excitatory and inhibitory areas. There are two main classes of complex cell. Both respond best to moving edges or slits of fixed width and precise orientation. One type of cell 'likes' a stimulus bar of a particular length, while the other type 'likes' end-stopped stimuli (these cells were previously called *hypercomplex cells*). The best stimuli for these cells requires not only

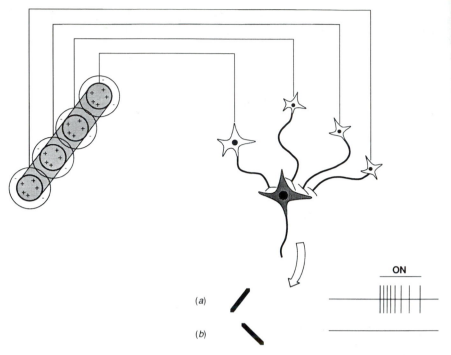

Fig. 5.1. A simple orientation selective cell (the grey cell in the diagram) behaves as if it receives input from several centre–surround antagonistic cells of the LGN (shown in white). Flashing a line with an orientation that stimulates more of the excitatory centres will stimulate the cell. Thus a bar oriented as in (a) will stimulate the cell, but one oriented as in (b) will not. (Redrawn from Hubel & Wiesel, 1962.)

a certain orientation but also a discontinuity, such as a line that stops, an angle or a corner.

Spatial frequency

Although many neurons in V1 do respond to lines and bars, the optimal stimulus is a sine-wave grating (De Valois, Albrecht & Thorell, 1978). A sine-wave grating looks like a series of fuzzy, unfocused parallel bars. Along any line perpendicular to the long axis of the grating, the brightness varies according to a sine-wave function. Sine-waves can vary in frequency and amplitude. The location of a point along a single cycle of a sine-wave is specified by its phase. The peak is at 0 degrees, the middle of the cycle is at 90 degrees, the trough is at 180 degrees and the middle of the cycle occurs again at 270 degrees. The

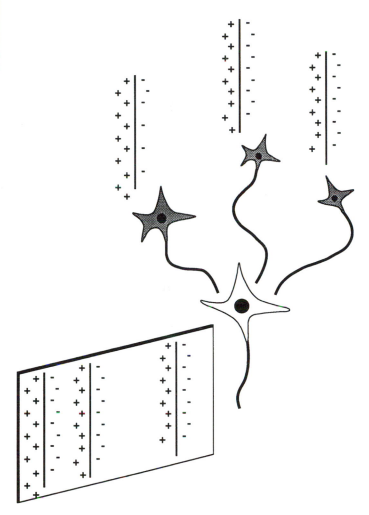

Fig. 5.2. The receptive field organisation of a complex cell in V1. These complex cells (shown in white in the diagram) are also responsive to lines of specific orientation and behave as if they receive an excitatory input from several simple cells (shown in grey) with oriented receptive fields. (Redrawn from Hubel & Wiesel, 1962.)

end of the cycle, 360 degrees, is the same as the beginning of the next one. A sine-wave is classified by its spatial frequency. Because the size of the image of a stimulus on the retina is dependent upon how close it is to the eye, visual angle is usually used instead of the physical distance between adjacent cycles. Therefore, the spatial frequency of a sine-wave grating is measured in cycles per degree of visual angle. Most neurons in the striate cortex respond best when a sine-wave grating of a particular spatial frequency is placed in the appropriate part of the visual field. For orientation-selective neurons, the grating must be aligned at the appropriate angle of orientation. In most cases, a neuron's receptive field is large enough to include between 1.5 and 3.5 cycles of the grating (De Valois, Thorell & Albrecht, 1985).

Texture

Recently a new class of neuron has been reported in V1, which seems to be responsive to texture (von der Heydt, Peterhans & Dürstler, 1992). They are unresponsive to single lines, bars, or edges, but they respond preferentially to a sine-wave or square-wave grating of a particular spatial frequency and orientation. These cells do not seem to be frequency analysers like those described above but, instead, respond to the 'texture' of the pattern. Natural surfaces have a rough texture, and these cells sensitive to texture and texture orientation can not only detect the presence of a surface but also its orientation. They may contribute to depth perception, for which texture gradients are an important cue.

Direction selectivity

Around 10–20% of complex cells in the upper layers of the striate cortex show a strong selectivity for the direction in which a stimulus is moving. Movement in one particular direction produces a strong response from the cell, but it is unresponsive to movement in other directions. The other complex cells do not show a marked direction preference. How are such direction selective cells wired up? In 1965, Horace Barlow and William Levick proposed a wiring plan for direction selective cells in the rabbit retina; this has been modified in order to explain the direction selectivity of the primate complex cells and is illustrated in Fig. 5.3 (Hubel, 1989). It is suggested that between simple and complex cells there are intermediate cells. These cells receive excitation from one simple cell and inhibition from a second simple cell via a second intermediate cell. The second intermediate cell has a receptive field that is immediately adjacent to that of the first. If a stimulus moves in the null direction then the first intermediate cell is excited by one of its inputs just as it is inhibited by the other, whose receptive field has just been crossed. The two effects cancel and the cell does not fire. If the stimulus moves in the opposite direction then the inhibition arrives too late to prevent the cell from firing.

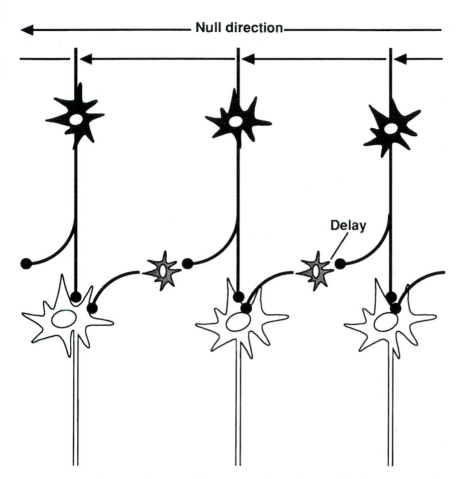

Null direction

Delay

Fig. 5.3. A schematic diagram of the proposed neural circuit for detecting motion. The synapses from black to grey are excitatory and from grey to white are inhibitory, The three white cells at the bottom of the figure are supposed to converge on a single master cell. (Redrawn from Hubel, 1989.)

If an observer views a grating moving in a single direction for several minutes, the observer's threshold to detect motion in this direction is increased by up to a factor of two. This is an example of selective adaptation, produced by fatiguing the cortical cells that are selective for movement in that direction (Hammond, Movat & Smith, 1985). The arrangement of cortical neurons for the detection of motion suggests a ready explanation for the phenomenon of motion aftereffect, sometimes called the waterfall illusion. If you view a stimulus moving in one direction for a period of time, such as a grating or a waterfall and then fixate a stationary object, it will appear to move in the opposite direction to the moving stimulus. This is because prolonged exposure to downward motion will fatigue or adapt the cells preferring downward motion. They will have virtually no spontaneous activity, while those preferring upward motion will have normal levels of spontaneous activity. This biased distribution of spontaneous activity produces a pattern of activity similar to that produced by actual upward movement and it is believed that this forms the neural basis of the motion aftereffect (Mather & Moulden, 1980).

Colour

As described in Chapter 4, staining for the mitochondrial enzyme cytochrome oxidase has shown a matrix of oval patches or blobs, each approximately $150 \times 200 \, \mu$m; each blob is centred on an ocular dominance column. Within an ocular dominance column, the degree of dominance varies. In the centre of a column, the dominance will be absolute and the neurons will receive input from only one eye. The blobs coincide with these centres of monocularity (Ts'o et al., 1990). The blobs contain colour-opponent and double colour-opponent cells, but there is a segregation in the different forms of colour opponency. Within a blob, the neurons will be either red/green opponent or blue/yellow opponent, the two forms of opponency are not mixed within a single blob (T'so & Gilbert, 1988). These findings suggest that individual blobs are dedicated to the processing of one colour-opponency system. Moreover, different types of blob are not equally represented in V1. There are more red and green cones in the retina than blue cones, and this is reflected in the proportions of the red/green opponent retinal ganglion cells to the blue/yellow ganglion cells (5:2). This difference is also reflected in the proportions of red/green blobs to blue/yellow blobs (3:1); the blue/yellow blobs seem to be clustered together. This suggests a non-uniform or patchy input of blue/yellow inputs into V1, which would be consistent with the annular organisation of blue cones in the retina.

The blobs often seem paired. The blob in a particular ocular dominance column is connected by *bridges* to a blob in a neighbouring ocular dominance column. In vertical section, this gives a ladder-like arrangement, with the bridges forming the rungs. Microelectrode recording and optical imaging studies using voltage-sensitive dyes have shown that these bridges contain colour-selective cells like those in the blobs (T'so & Roe, 1995). Bridges have been found connecting blobs of different colour opponency. In this case, the responses of neurons in the bridges are neither red/green nor blue/yellow but are mixed in spectral selectivity.

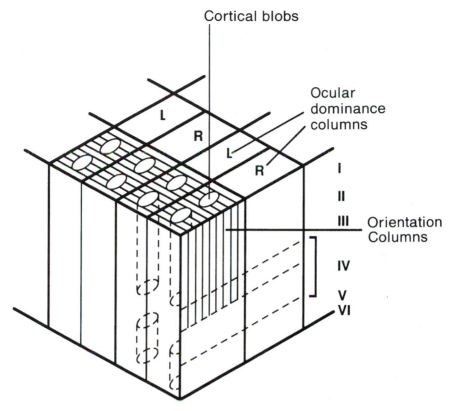

Fig. 5.4. A schematic diagram of the proposed organisation of V1, often called the 'ice-cube' model. The primary visual cortex is divided into modules containing two blobs and their intrablob region. The cortex is divided into ocular dominance columns; running perpendicular to these are orientation columns.

Modular organisation

Different visual areas seem to be subdivided into processing *modules* or *super columns*, which receive information from other modules, perform some calculation and then pass it on to other modules. V1 has been suggested to contain 2500 modules each 0.5×0.7 mm and each containing 150 000 neurons. The neurons in each module are concerned with the analysis of a particular portion of the visual field. The modules consist of two units, each centred around a blob (Fig. 5.4). The neurons in layer $4C\alpha$ and $4C\beta$ are monocular. Each unit receives input from the opposing eye. The two units, however, exchange information and 80% of the neurons in the other layers

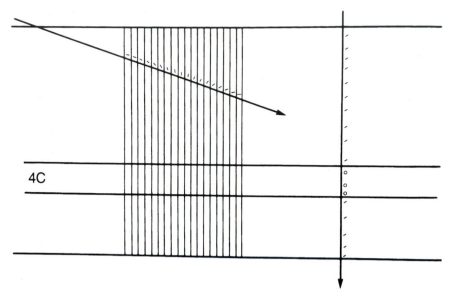

Fig. 5.5. A schematic diagram of the orientation columns in monkey V1. Two microelectrode penetrations are illustrated, one vertical and one oblique. In the vertical penetration, the orientation is clearly defined and constant, from cell to cell, in layers above and below layer 4c. In layer 4c the cells have centre–surround receptive fields, without orientation preferences. In an oblique penetration, there is a systematic variation in orientation, either clockwise or counter clockwise, in steps of 10 degrees. (Redrawn with permission from Hubel & Wiesel, 1977. Copyright (1977) Royal Society.)

are binocular. As mentioned, neurons in the blobs are sensitive to wavelength, and neurons in the interblob region are sensitive to orientation and, in the case of the complex cells, movement as well. Therefore, each module contains neurons sensitive to wavelength, movement and lines or edges of particular orientations within a specific portion of the visual field. Within a module the response characteristics of a cell are arranged systematically. If a neuron in one position responds best to a line oriented at 45 degrees, then a neuron a short distance away will respond best to a line oriented at 50 degrees. Each $25\,\mu$m of lateral movement encounters a neuron that responds to a line rotated by 10 degrees (Fig. 5.5). Travelling across both halves of the module there are two 180 degree rotations in the orientations of lines that best stimulate the neurons.

The orderly progression of stimulus orientation in columns across the cortex, as shown in the traditional 'ice cube' model, seems to be an idealised version of cortical organisation. The microelectrode experiments described above can only be used to map small areas of V1, but a new technique for simultaneously mapping activity across the cortex has been developed by Gary Blasdel (Blasdel & Salma, 1986). Blasdel removed that part of the monkey's skull covering the cortex of V1 and replaced it with a glass window. He then injected a voltage-sensitive dye onto the surface of the cortex. The colour of this dye changes with the strength of the electrical current passing through it, and so it can be used to analyse patterns of electrical activity in the striate cortex. If the visual stimuli used are bars of different orientations, it is possible to map the distribution of the orientation columns; if the stimuli are presented monocularly, it is possible to map the ocular dominance columns. This technique has shown that although there are regions of cortex where a smooth gradation of orientation occurs there are also discontinuities and distortions in the mapping, where columns with markedly different preferences are sometimes neighbours. This makes sense if you think about it. The stimulus can vary in orientation, ocular dominance and two dimensions of retinal position (up or down and from side to side) and to map this onto an irregular-shaped piece of two-dimensional cortex will require the occasional discontinuity to allow most of the retinotopic relationships to be preserved (Young, 1993a).

Key points

1. In cats and monkeys, the left and right eyes supply alternate layers of the LGN. The cells from one LGN layer project to groups of target cells in layer 4C, separate from those supplied by the other eye. These groups of cells form alternating stripes or bands in layer 4C. Above and below this layer, most cells are driven by both eyes, although one eye is usually dominant. Hubel and Wiesel termed these blocks of cells ocular dominance columns.

2. There seem to be two broad categories of neurons in V1, termed simple and complex neurons. In addition there is a class of cell found exclusively in layer 4C (where most LGN fibres terminate), that has a concentric centre–surround receptive field like those of the LGN and the retinal ganglion cells.

3. Around 10–20% of complex cells in the upper layers of the striate cortex show a strong selectivity for the direction in which a stimulus is moving. Movement in one particular direction produces a strong response from the cell, but it is unresponsive to movement in other directions.

4. Staining for the mitochondrial enzyme cytochrome oxidase has shown a

matrix of blobs on the surface of V1, approximately $150 \times 200 \, \mu$m each. Each blob is centred on an ocular dominance column. The blobs contain colour-opponent and double colour-opponent cells, but there is a segregation in the different forms of colour opponency. Within a blob, the neurons will be either red/green opponent or blue/yellow opponent. The two types of blob are present in unequal proportions (three red/green blobs to one blue/yellow blob), and the blue/yellow blobs seem to be clustered together. The blobs often seem paired. The blob in a particular ocular dominance column is connected by bridges to a blob in a neighbouring ocular dominance column.

5. Different visual areas seem to be subdivided into processing modules or super columns. V1 has been suggested to contain 2500 modules each 0.5×0.7 mm and each containing 150 000 neurons. Each module contains neurons sensitive to wavelength, movement and lines or edges of particular orientations within a specific portion of the visual field. Within a module, the response characteristics of a cell were proposed to be arranged systematically, such as in the 'ice cube' model. However, optical imaging studies have shown that although there are regions of cortex where a smooth gradation of orientation occurs there are also discontinuities and distortions in the mapping, where columns with markedly different preferences are sometimes neighbours.

6

Visual development: an activity-dependent process

Variations on a theme

The development of the visual system is under the control of both genetic and environmental factors. The connections are refined and cut to fit on the basis of neural activity that is constantly flickering through the visual system from the retina. Following birth, it is environmental stimulation that elicits neural activity in the visual system. Cells in the retina, LGN and V1 of newborn, visually naive monkeys and kittens have receptive field and response properties very much like those of the adults. However, there are differences in their visual systems, such as in layer 4 of V1 where the projections from the LGN terminate. At birth, the cells in layer 4 are driven by both eyes, as projections from the LGN spread over a wide region of layer 4, whereas in the adult a layer 4 cell is driven by either eye but not by both. The adult pattern of ocular dominance columns in layer 4 is established over the first six weeks of life, when the LGN axons retract to establish separate, alternating zones in layer 4 that are supplied exclusively by one eye or the other (Fig. 6.1).

In early life, the connections of neurons in the visual system are susceptible to change and can be irreversibly affected by unbalanced neural activity passing through them. For example, closure of the lids of one eye during the first three months of life leads to blindness in that eye. This is not because the eye no longer functions properly, but because the neurons in the visual cortex no longer respond to the signals the eye sends to them. Lid closure in adult animals has no such effect. It seems that for the visual system to be correctly wired up, it must receive stimulation from the eyes to guide its development

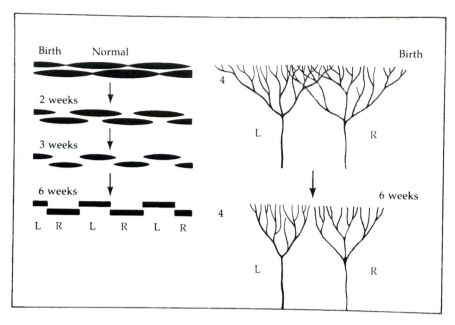

Fig. 6.1. A schematic representation of the retraction of cat LGN axons which terminate on layer 4 of the visual cortex during the first six weeks of life. The overlap of the inputs from the right (R) and the left (L) eye present at birth gradually become segregated into separate clusters corresponding to the ocular dominance columns. (Redrawn from Nicholls, Martin & Wallace, 1992.)

and allow connections to be strengthened or weakened depending on the activity in the system. The most favoured theory for the mechanism under-lying neural plasticity in adult animals was proposed back in the 1940s by Hebb. He suggested a coincidence detection rule such that when two cells are simultaneously active, the synapse connecting them is strengthened (Hebb, 1949). The discovery of a putative cellular substrate for learning long-term potentiation (LTP), by Lømo in 1966 has resulted in a veritable deluge of studies. This work has, very largely, been centred on the hippocampus, an important area for learning and memory. In the hippocampus, LTP is characterised by an abrupt and sustained increase in the efficiency of synaptic transmission brought on by a brief high-frequency stimulus. It may persist in the *in vitro* hippocampal slice for hours and in a freely moving animal for days (Bliss & Collingridge, 1993). Although LTP does seem to be the most likely candidate for the mechanism of activity-dependent synaptic plasticity, it con-tinues to be extraordinarily difficult to determine exactly how this synapse strengthening comes about. One generally agreed feature is that N-methyl-D-aspartate (NMDA) receptors, a subtype of glutamate receptor, mediate the entry of Ca^{2+} in CA1 of the hippocampus and thus induce LTP, although NMDA receptors are not necessarily involved in LTP at other sites. The NMDA receptors open in the presence of L-glutamate when the post-synaptic membrane is depolarised sufficiently to expel the channel blocker Mg^{2+}. Much of the current debate concerns the site that controls LTP expression: is it presynaptic or postsynaptic or is control dependent on the specific experimental conditions? In spite of this continuing tussle, the evidence for LTP as a general model of synaptic plasticity in the adult brain is increasing. But what of the plasticity involved in the developing brain – is there a common mechanism? This chapter will examine the evidence for changes in the visual system with changes in visual input, and the possible mechanisms that might mediate these changes.

Monocular or binocular deprivation

The segregation of the LGN axons to form ocular dominance columns seems to be dependent on balanced activity from the two eyes. If this activity is interrupted, and the balance between the two eyes is altered, then the result is a series of changes in the organisation of the visual system. One rather drastic way of altering the balance of activity is to close one eye in a developing animal. Rearing kittens with one eye sutured (*monocular deprivation*) causes a series of changes throughout the visual system and drastically reduces the perceptual capabilities of the eye that has been sutured during early development. In the LGN, neurons connected to the deprived eye were reduced in size by 40% relative to the neurons connected to the other eye (Wiesel & Hubel, 1963). Further, studies on the terminal fields of the LGN cells in layer 4 showed that LGN axons connected to the deprived eye occupied less than 20% of the cortical area, and the other non-deprived eye had expanded its representation to cover more than 80% of the thalamic recipient zone (LeVay, Stryker & Shatz, 1978). Single-unit recording studies have shown that stimuli presented through the formerly deprived eye failed to influence the majority of cells in the striate visual cortex (Fig. 6.2). The undeprived eye becomes the primary effective route for visual stimuli.

Under conditions of dark rearing (*binocular deprivation*), the organisation of the visual system and the selectivity of the cells initially continues to develop, despite the lack of visual stimuli (Buisseret & Imbert, 1976). When both eyes

(a)

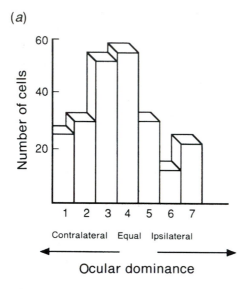

(b)

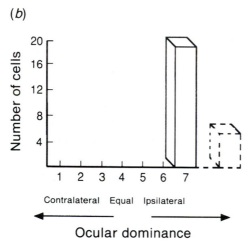

Fig. 6.2. Ocular dominance histograms in cells, recorded from the visual cortex of cats. (a) Recordings for 223 cells of adult cats. Cells in groups 1 and 7 of the histogram are driven by one eye only (ipsilateral or contralateral). All the other cells have inputs from both eyes. In groups 2, 3, 5 and 6, the input from one eye is dominant. In group 4, both eyes have a roughly equal influence. (b) Recordings from 25 cells of a 2.5-month-old kitten that was reared with its right eye occluded until the time of the experiment. The dashed bar on the right indicates that five cells did not respond to the stimulation of either eye. The solid bar indicates that all 20 cells that were responsive to stimulation responded only to the eye that was opened during rearing. (Redrawn from Wiesel & Hubel, 1963.)

are closed in newborn monkeys for 17 days or longer, most cortical cells (such as the simple and complex cells) respond largely as normal to visual stimuli (Daw *et al.*, 1983). The organisation of layer 4 seems to be normal and in other layers most cortical cells are stimulated by both eyes. The major difference is that a large proportion of cells could not be driven binocularly, and some spontaneously driven cells could not be driven at all while others were less tightly tuned to stimulus orientation. Binocular deprivation in kittens leads to similar results except that more cortical cells continue to be binocularly driven (Wiesel & Hubel, 1965). Longer visual deprivation (three months or more) leads to a more marked effect. The visual cells become weakly responsive or totally unresponsive to visual stimuli, and the weakly responding cells lack orientation, direction and stereo selectivity (Sherk & Stryker, 1976; Pettigrew, 1974). It seems that some of the results of monocular deprivation can be prevented or reduced by binocular deprivation. It may be that the two eyes are *competing* for representation in the cortex, and with one eye closed the contest becomes unequal.

What, then, is the physiological basis for this ocular dominance shift associated with monocular deprivation? Such a shift can be prevented by modifying neuromodulator and neurotransmitter functions in the cortex (e.g. Shaw & Cynader, 1984; Bear & Singer, 1986; Reiter & Stryker, 1988), for example, by the infusion of glutamate into the cortex for a two-week period during monocular deprivation. Control recordings during the infusion period show that cortical neurons in general fail to respond well to visual stimuli from either eye during the infusion period. The lack of ocular dominance modification seems to be the result of the reduced ability of the cortical cells to respond to the unbalanced LGN afferent input. Effective inputs representing the two eyes are greater than that of the deprived eye. It seems that, although changes associated with monocular deprivation have been found at the level of retino-geniculate terminals, in the LGN cell bodies, in the LGN terminal and in cortical cells' responses, the primary site of binocular competition is cortical, and other changes in the visual system are secondary to the primary cortical competition.

This change in ocular dominance, as in all major rewiring in the visual system, occurs during a limited period following birth, often called the *critical* or *sensitive period*. It seems that the visual system is only capable of rewiring itself in this small temporal window and can do very little more once this opportunity has elapsed. For kittens, deprivation for only 3 days between the fourth and fifth week causes a large change in the pattern of ocular dominance (Hubel & Wiesel, 1970). If deprivation was started later than the eight week, smaller effects were observed, until even long periods of deprivation at four

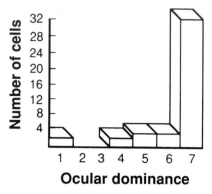

Fig. 6.3. Ocular dominance histogram of a kitten that had one eye occluded for 24 hours following four weeks of normal vision. (Redrawn from Olson & Freeman, 1975.)

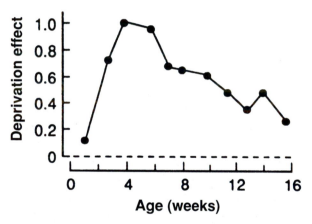

Fig. 6.4. Profile of the sensitive period for monocular deprivation in kittens. As can be seen, the most sensitive period is at four to six weeks, but monocular deprivation can cause substantial effects as long as four months after birth. (Redrawn from Olson & Freeman, 1980.)

months caused no effect (Fig. 6.3). Hubel and Wiesel concluded that the critical period for susceptibility to monocular deprivation begins in the fourth week and extends to about four months of age. Deprivation does not have to be long to cause large effects if it occurs during the critical period. Occluding one of the eyes of a four week-old kitten for a single day causes a large effect on the pattern of ocular dominance (Olson & Freeman, 1975) (Fig. 6.4). However, it seems that the critical period is not fixed (Cynader, 1983). If cats are reared in the dark until long after the end of the chronologically defined critical period and only then brought into the light for monocular deprivation, this deprivation can still produce marked effects on cortical ocular dominance. Dark-reared cats seem to undergo a new critical period in the first few weeks after they are brought into the light. So, not only is the rewiring activity dependent but so is its initiation.

A similar situation to that described above in non-human mammals for monocular deprivation may also occur in human subjects. There is increasing evidence that *amblyopia* (a large reduction in the visual acuity in one eye) may sometimes occur in humans who as young children had reduced use of one eye because of patching following an eye operation. This evidence has been provided by investigating the histories of 19 patients with amblyopia and finding that they all had their amblyopic (low visual acuity) eye closed in early life, following an eye operation, with most of the closures occurring within the first year of life (Awaya *et al.*, 1973). This type of amblyopia is called *stimulus-deprivation amblyopia*.

Image misalignment and binocularity

The changes in the ocular dominance columns are merely the most obvious effect of changes in the balance of neural input into the visual cortex. The majority of cells in the normal visual cortex are binocular, and during postnatal development, when the ocular dominance columns are being established, the connections to individual cells from both eyes are also being refined. Unsurprisingly, monocular deprivation leads to most cells in the visual cortex being monocular. Under normal conditions, the input to a cell from the two eyes is from corresponding areas of the retina. Misalignment of the images in the two eyes can be accomplished either by cutting the eye muscles or by fitting the animal with a helmet that contains small optical prisms. This disruption does not alter the absolute magnitude of activity. Under these conditions, most cells can only be driven monocularly, rather than binocularly as in normal animals, and the ocular dominance columns seem more sharply delineated (Lowel & Singer, 1993). Whereas 80% of cortical cells in normal cats are binocular, only 20% of the cells in cats with cut eye muscles respond to the stimulation of both eyes (Hubel & Wiesel, 1965). Similarly, 70% of cortical cells in monkeys are binocular, but less than 10% of cells are binocular in a monkey that has worn a prism-helmet for 60 days (Crawford & von Noorden, 1980). These neurological changes translate into striking behavioural effects. For example, prism-reared monkeys are unable to detect depth in random-dot stereograms, suggesting that they have lost the ability to use binocular disparity to perceive depth (Crawford *et al.*, 1984).

It appears that it is not just the magnitude or balance of neural activity that is important, but also the temporal pattern of this activity. This hypothesis is supported by experiments in which the retina was deactivated with tetrodotoxin, and the optic nerve was directly stimulated. This allowed the temporal relationship of the neural activity from the two eyes to be directly controlled. Many more cortical cells were found to be monocular under a regimen of separate stimulation through the two optic nerves than were found with simultaneous stimulation (Stryker and Strickland, 1984). This synchronised activity of the two inputs to the same cell could be used in a Hebbian process for strengthening synapses from both inputs. The mechanism of this strengthening could be a form of LTP called 'associative LTP', in which the paired activity of two inputs to a cell results in the strengthening of both inputs. Uncorrelated activity from the two eyes seems to lead to a weakening and possible elimination of synapses in the visual cortex. This is another form of neural plasticity, long-term depression (LTD).

Image misalignment in humans

Some people have an imbalance in the eye muscles that upsets the co-ordination between their two eyes. This condition is called *strabismus*. Just as cutting the eye muscles in experimental animals causes a loss of cortical cells that respond to stimulation of both eyes, there seems to be a similar lack of binocularly driven cells in people who had strabismus as young children. Strabismus can be corrected by a muscle operation that restores the balance between the two eyes. However, if this operation is not performed until the child is four-five years of age, a loss of binocularly driven cells seems to occur. This can be measured by the *tilt aftereffect* (Fig. 6.5), because of the phenomenon of interocular transfer. If an observer looks at the adapting lines with one eye and then looks at the test lines with the other eye, the aftereffect will transfer between the eyes. This transfer, which is about 60–70 % as strong as the effect that occurs if the adaptation and test lines are viewed with the same eye, indicates that information from one eye must be shared with the other. The degree of transfer can be used to assess the state of binocularly driven cells. When surgery is carried out early in life, interocular transfer is high, indicating good binocular function, but if the surgery is delayed, interocular transfer is poor, indicating poor binocular function. The critical period for binocular development in humans seems to begin during the first year of life, reaches a

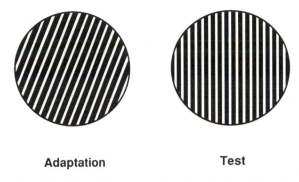

Adaptation   **Test**

Fig. 6.5. Stimuli for measuring the tilt aftereffect. If you stare at the adaptation pattern on the left for 60 s and then turn your gaze on to the test pattern on the right, you see the test lines as tilted. This is the tilt aftereffect.

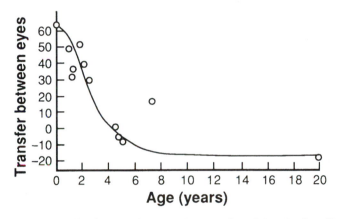

Fig. 6.6. The degree of interocular transfer of the tilt aftereffect as a function of the age at which surgery was performed to correct strabismus (after Banks, Aslin & Letson, 1975).

peak during the second year, and decreases by four to eight years (Banks, Aslin & Letson, 1975) (Fig. 6.6).

Selective rearing: manipulating the environment

Another way of altering the visual input is to raise animals in a tightly controlled visual environment, dominated by a certain visual stimulus and deficient in others. This does not alter the balance of activity between the eyes, but does alter the pattern of activity produced by each eye. These experiments have usually been carried out either by placing infant animals in an environment containing stripes of only one orientation (e.g. Blakemore & Cooper, 1970) or by fitting infant animals with goggles that present vertical stripes to one eye and horizontal stripes to the other (e.g. Hirsch & Spinelli, 1970).

Blakemore and Cooper kept kittens in the dark from birth to two weeks of age and then placed them in a large vertical tube for 5 hours every day. For the rest of the day they remained in the dark. The inner surface of the tube was covered with either horizontal or vertical stripes. The kittens sat on a plexiglass floor and the tube extended above and below them to ensure there were no visible corners or edges in their environment other than the stripes on the sides of the tube. The kittens wore neck ruffs to prevent them altering the orientation of the stripes by turning their heads. After 5 months, the selective rearing was stopped and the kittens remained in the dark except for brief periods when their vision was tested. The kittens displayed a number of defects in their visual behaviour. Their head movements were jerky when following moving objects, they tried to touch distant objects and often bumped into things. Most important of all, they seemed to be blind to stripes orthogonal to the orientation of the environment in which they were reared. Following these behavioural tests, Blakemore and Cooper recorded from cells in the visual cortex to determine the optimum stimulus orientation for different cells. Most of the cells of the 'horizontally reared' cats responded primarily to horizontal stimuli and none at all responded to vertical stimuli. The opposite is true of the 'vertically reared' cats. These results have been confirmed by subsequent experiments (Muir & Mitchell, 1975). The results of Hirsch and Spinelli's experiments (1970) using goggles showed the same pattern of effects. In single-cell recording experiments, they found few cells in the visual cortex where the preferred orientation deviated from the orientation of the environmental stimulus by more than 5–10 degrees.

A single hour of exposure in a striped tube can drastically alter the preferred orientation of cells in the visual cortex. Blakemore and Mitchell (1973) kept a kitten in the dark until 28 days of age, placed the kitten in a tube with vertical stripes for a single hour and then kept the kitten in the dark until they recorded from its visual cortex at 42 days of age. As in kittens exposed to vertical stripes for much longer periods of time, most cells responded best to vertical or near-vertical orientations.

A number of different types of environment have been used to explore cortical plasticity further, such as moving white spots, random arrays of point sources of light and stripes moving in one particular direction (e.g. Van Sluyters & Blakemore, 1973; Pettigrew & Freeman, 1973). In each case, the majority of cortical cells responded to the stimuli that were present in their environment and responded very weakly to anything else. An interesting example of selective rearing is shown in cats reared under conditions of stroboscopic illumination, where continuous retinal movement is prevented. This results in a deficit in the direction selectivity of the cortical cells, which is expressed behaviourally as a deficit in motion perception (Cynader & Cherneneko, 1976). The development of other cortical cell properties, such as orientation and stereoselectivity, is unaffected.

It seems that during the critical period a number of changes are made to the wiring of the visual system, and this fine tuning is activity dependent. Without neural activity to stimulate and alter the strength of synaptic connections, the normal response properties of visual cells will not develop.

Impoverished visual input in humans

The selective rearing experiments have been used as a model of a condition in humans called *astigmatism*. This is caused by a distortion in the cornea, which results in an image that is out of focus either in the horizontal or the vertical orientation. A person who has an astigmatism at an early age is exposed to an environment in which lines in one orientation are imaged sharply on the retina, but lines 90 degrees from this orientation are out of focus. Freeman and Pettigrew (1973) showed that cats reared with an artificial astigmatism, created by wearing a mask containing astigmatic lenses, develop cortical cells that favour whichever orientation is in sharp focus during rearing. This result in cat vision resembles a condition known as *meridional amblyopia* in humans. People whose vision is not corrected very soon after birth seem to show the same perceptual changes as animals reared in a selective environment or with goggles. As a result, even if the optical errors are subsequently corrected, the subject's vision will still be poor as he or she will not have the cortical machinery to process the new information available to the visual system.

The critical period

LTP and LTD are closely linked to the function of NMDA receptors. These receptors are found in the visual cortex of both cats and kittens (Fox, Sato & Daw, 1989), and the blocking of these receptors prevents the ocular dominance shift that occurs after monocular deprivation (Bear *et al.*, 1990). These studies provide strong circumstantial evidence for a role for LTP and LTD in activity-dependent refinement of the visual cortex. Whilst most experiments on the development of the visual cortex have used monkeys and cats, the same effects can also be found in rat visual cortex (Fagiolini *et al.*, 1994), and in the last few years the hapless rat has formed the basis of brain slice preparations to investigate neural plasticity. The occurrence of LTP in the adult rat visual cortex was first reported some years ago (Artola & Singer, 1987), but, recently, both LTP and LTD were also found to occur in the white matter of layer 3 of visual cortex in post-natal rats (Kirkwood & Bear, 1994; Kirkwood, Lee & Bear, 1995). Most interestingly of all, a form of LTP was reported that only occurs during the critical period. Moreover, when the critical period is shifted by binocular deprivation, the occurrence of this form of LTP shifts with it, so the two are always in register.

Such a coincidence between LTP occurrence and a critical period is not confined to the visual system. In the rat somatosensory cortex, connections can be radically altered if the input from the sensory vibrissae around the rat's nose and mouth is manipulated. In this case, the critical period has no overlap with that regulating input to the visual cortex but is much earlier and seems to be confined to the first post-natal week (Schlaggar, Fox & O'Leary, 1993). Crair and Malenka (1995) have found a form of LTP in the somatosensory cortex that can only be induced during this first week. In addition, they present evidence for the involvement of NMDA receptors in this critical period LTP, which adds to the likelihood that the LTP found in the adult brain is the same, or very similar, to that involved in the major construction that is carried out in the post-natal developing brain. Unlike the situation in the adult brain, however, these NMDA receptors must undergo some chemical or structural change correlated with the decline in LTP with the end of the critical period. The molecular composition of NMDA receptors can change during development, although the trigger for this is, as yet, unknown (Williams *et al.*, 1993).

What we see shapes how we see it

The development of the visual system combines features of both a hard-wired network and a self-organising neural net. The basic structure is pre-determined and is largely unaffected by the neural activity passing through it. However, for all the complex connections to be precisely specified in advance would be an epic task, and the opportunity for error during development would be immense. Therefore, the fine tuning of the connections, including the target cell for a particular LGN afferent as well as the balance and weighting of the synapses, is an activity-dependent process mediated by specific forms of LTP and LTD. As a result, our visual experience in the period immediately following birth is vitally important in shaping the functional organisation of visual system, and an imbalance in visual stimulus will be mirrored by an imbalance in the visual system's organisation.

Key points

1. The development of the visual system is dependent upon balanced neural activity from the eyes during a critical period early in post-natal development. Disruption of this activity through monocular deprivation or a controlled environment disrupts the organisation of the visual system.
2. Monocular deprivation in young mammals leads to the visual cortex becoming unresponsive to the covered eye. A similar situation can be found in children who have had one eye patched. In humans, this condition is called amblyopia.
3. Binocular deprivation has a less dramatic effect initially, although longer visual deprivation (three months or more) leads to visual cells becoming weakly responsive or totally unresponsive to visual stimuli; the weakly responding cells lack orientation, direction and stereoselectivity.
4. Some of the results of monocular deprivation can be prevented or reduced by binocular deprivation or by the infusion of neurotransmitter blockers; it seems that the two eyes are competing for representation in the cortex, with one eye closed the contest becomes unequal.
5. If there is a misalignment of the images in the two eyes during early development, the proportion of visual neurons that are binocular are drastically reduced. These neurological changes translate into striking behavioural effects, suggesting that such animals have lost the ability to use binocular disparity to perceive depth.
6. A similar lack of binocularly driven cells can also be found in human subjects who during early childhood have had an imbalance in the eye muscles that upsets the co-ordination between the two eyes. This condition is called strabismus.
7. If an animal is raised in a controlled environment where it only sees certain stimuli, such as only horizontal lines, it will be behaviourally and neurophysiologically unresponsive to lines of other orientations.
8. The selective-rearing experiments have been used as a model of a condition in humans called *astigmatism*. This is caused by a distortion in the

cornea which results in an image that is out of focus either in the horizontal or the vertical orientation.

9. The neural basis of cortical plasticity is two mechanisms called long-term potentiation (LTP) and long-term depression (LTD). Both forms are found in the visual cortex and one type of LTP is only found during the critical period.

7

Colour constancy

The colour constancy problem

One of the most important functions of the visual system is to be able to recognise an object under a variety of different viewing conditions. For this to be achieved, the stimulus features that make up that object must appear constant under these conditions. If stimulus parameters do not form a reliable 'label' for an object under different conditions, they are considerably devalued in their use to the visual system. For example, if we perceive a square shape on a video screen and the area it covers increases or decreases, we experience a sense of movement. The square seems to get closer or further away. The visual system assumes that the size of the square will not change, so that changes in its apparent size will signal changes in its relative distance from us. This is called *object constancy*. This is a sensible assumption as, under normal conditions, objects seldom change in size. Another example is *lightness constancy*. Over the course of a normal day, light levels change significantly, but the apparent lightness of an object will change very little. The visual system scales its measure of lightness to the rest of the environment, so that apparent lightness of an object will appear constant relative to its surroundings. A similar problem exists with the perception of colour. Over the space of a day, the spectral content of daylight changes significantly. This means that the spectral content of light reflected from an object changes too. One might expect that objects and surfaces acquire their colour as a result of the dominant wavelength of the light reflected from them; thus a red object looks red because it reflects more long-wave (red) light. However, surfaces and objects retain their colour in spite of wide ranging changes in the wavelength and energy composition of the light reflected from them. This is called *colour constancy* and is displayed not only by humans and primates but also by a wide range of species from goldfish to honeybees. So it seems there is no pre-specified wavelength composition that leads to a colour and to that colour alone. If colours did change with every change in illumination then they would lose their significance as a biological signalling mechanism since that object could no longer be reliably identified by its colour.

The Land Mondrian experiments

Some of the most important and influential studies on colour constancy were made by Edwin Herbert Land (1909–91). Land was a Harvard University drop-out who went on to become one of the most successful entrepreneurs in America (Mollon, 1991). He developed a method for producing large sheets of artificial polariser and in 1937 founded the Polaroid Corporation to market his invention. Polaroid filters, for visible and infra-red light, were soon being used in cameras and sunglasses, and in wartime range-finders and night-adaptation goggles. This development was followed up in 1948 with an instant camera, which could produce a picture in 60 s, and Land and his company became very rich. However, for the last 35 years of his life, Land's chief obsession was with colour and colour constancy. As part of his experiments, he had observers view a multicoloured display made of patches of paper of different colour pasted together (Land, 1964). This display was called a *Colour Mondrian*, from the resemblance it bore to the paintings of the Dutch artist Piet Mondrian. The rectangles and squares composing the screen were of different shapes and sizes, thus creating an abstract scene with no recognisable objects to control for factors such as learning and memory. No patch was surrounded by another of a single colour and the patches surrounding another patch differed in colour. This was to control for factors such as induced colours and colour contrast. The patches were made of matt papers that reflected a constant amount of light in all directions. As a result, the display could be viewed from any angle without affecting the outcome of the experiment.

The display was illuminated by three projectors, each equipped with a rheostat that allowed the intensity of the light coming from the projector to be changed. The first projector had a filter so that it only passed red light, the second projector only passed green light and the third projector only passed blue light. The intensity of light produced by each projector was measured using a telephotometer, so the relative amounts of the three wavelengths in the illumination could be calculated.

In one experiment, the intensity of light reflected from a green patch was set so that it reflected 60 units of red light, 30 units of green light and 10 units of blue light. Test subjects reported the green patch as being green in colour even though it reflected twice as much red as green light, and more red light than green and blue light put together. So this is a clear example of the perceived colour of the patch not corresponding to the colour of the predominant wavelength reflected from it.

This experiment was repeated but under slightly different conditions. The subject still observed the same patch, illuminated by the same light, but this time the patch was viewed in isolation. The surrounding colour patches were not visible. This is called the *void viewing condition*. In this case, the perceived colour of the patch corresponded to the wavelength composition of the light reflected from it. If the surround was then slowly brought into view, the colour of the patch was immediately reported to be green. This suggests that the perceived colour of the patch was determined not only by the wavelength composition of the light reflected from it, but also by the wavelength composition of the light reflected from the surrounding surfaces. If the position of the green patch was changed within the Mondrian, so that the surrounding patches were different, the perceived colour remained the same. This suggested that the relationship between the perceived colour and the wavelength composition of the patch and its surrounding patch or patches was not a simple one.

Reflectance and lightness: the search for constancy in a changing world

To construct a representation of colour that is constant with changes in the spectral illumination of a surface, the visual system must find some aspect of the stimulus that does not change. One physical constant of a surface that does not change is its *reflectance*. For example, a red surface will have a high reflectance for red light, and a low reflectance for green and blue light. If the intensity of the light incident upon the object changes, the proportions of red, green and blue light reflected from the object will not (Fig. 7.1). Therefore, the visual system must ignore the information related to light intensities and concentrate purely on relative reflectance.

One way of doing this is to compare the reflectance of different surfaces for light of the same wavelength. So, for example, consider two surfaces, a red and a green one. The red surface will have a high reflectance for red light and so reflect a high proportion of red light. The green surface will have a low reflectance for red light and, therefore, only a small proportion of red light

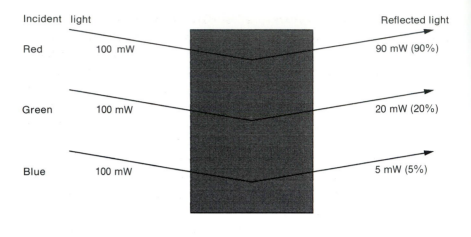

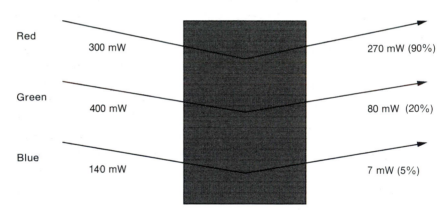

Fig. 7.1. The reflectance of a surface for light of a given wavelength is its efficiency for reflecting light of that wavelength, expressed as the percentage of the incident light of that wavelength which it reflects. The reflectance never changes, although the amounts incident on and relected from the surface change continually. The surface shown here reflects 90%, 20% and 5%, respectively, of red, green and blue light, irrespective of the intensity of the illuminating light. (Modified from Zeki, 1993.)

will be reflected from it. So if the patches are illuminated by a red light, the red patch will always appear lighter regardless of the intensity of the red light. Therefore, the biological correlate of reflectance is *lightness* (Zeki, 1993). By determining the efficiency of different surfaces in a scene for reflecting light of a given wavelength, the brain builds a lightness record of the scene for that particular wavelength.

When an entire scene is viewed, each surface will have a different lightness at every wavelength depending upon its efficiency for reflecting light of that wavelength. The record of that scene in terms of areas that are lighter or darker is called its *lightness record* (Zeki, 1993). In ordinary daylight, as in most light sources, there is a mixture of wavelengths, and each set of wavelengths will produce a separate lightness record. Land's *retinex theory* (the name is derived from retina and cortex) proposes that in the visual system the lightness records obtained simultaneously at three different wavelengths are compared in order to construct the colour of a surface (Land 1964; 1983). This comparison will be unrelated to the wavelength composition of the illuminating light and, therefore, will not be affected by the relative intensity of the lights of different wavelengths.

The colour that we perceive is, therefore, the end product of two comparisons: the comparison of the reflectance of different surfaces for light of the same wavelength (generating the lightness record of the scene for that wavelength), and the comparison of the three lightness records of the scene for the different wavelengths (generating the colour). Colour, therefore, is a comparison of comparisons (Zeki, 1992; 1993). When the wavelength composition of the light illuminating a surface changes, the intensities of light reflected from all the surfaces in the display will change, but the comparisons will remain the same because the reflectances do not themselves change.

Land has suggested an algorithm for generating these comparisons (Land, 1983). In it the logarithm of the ratio of the light of a given wavelength reflected from a surface (the numerator), and the average of light of the same wavelength reflected from its surround (the denominator) is taken. This constitutes a designator at that wavelength. The process is done independently three times for the three wavelengths.

The biological basis of colour constancy

Individual cells in the blobs of monkey V1 have small receptive fields and are responsive to certain wavelengths. For example, a cell might be responsive only to red light and be unresponsive to light of other wavelengths and to white light. If the red area of the Mondrian display is placed in this cell's receptive field and is illuminated with the standard triplet of energies (60 units red light, 30 units green light and 10 units blue light), the cell responds vigorously. The cell continues to respond vigorously if the green patch of the Mondrian display is moved into the cell's receptive field, even though the patch appears green to a human observer. The cell in V1 is responding to the fact that the primary wavelength of the reflected light is red. If the patches are seen in the void condition, the cell responds as it did before, although the human subject perceives a change in colour between the two conditions. The cell in V1 is responding not to the perceived colour but to the wavelength composition of a surface. There are two other types of colour-sensitive cell in V1: opponent and double-opponent cells. These also show a response only to the spectral composition of the light reflected from a patch.

In the retinex theory, before the colours are generated the lightnesses must be generated, since it is the comparison of the three lightness records that results in colour. A basic pre–condition for a neural candidate for this system is a lightness-differencing stage, in which a cell has centre–surround organisation in which the centre is excited by light of a particular wavelength and the surround is inhibited by light of the same wavelength (Zeki, 1993). The first such cells encountered in the primate visual system are the double-opponent cells of V1 (Zeki, 1983). They have spatially antagonistic centres and surrounds and give an ON response to red light and an OFF response to green light in the centre, and a response of the opposite polarity in the surround (Zeki, 1993). These cells are capable of detecting differences in lightness across the boundaries of their opponent fields for the same wavelength. However, their receptive fields are very small, and so these cells can only undertake this computation for a small part of the visual field.

Wavelength-selective cells whose responses seem to correlate with the human perception of colours have been reported in V4 (Zeki, 1983). The receptive fields of these visual cells are much larger than in V1. The red patch of a Mondrian was placed in the receptive field of a V4 cell responsive to red light, and the patch was then illuminated with the three projectors as for V1. The cell gave a good response. If areas of other colour, such as green, blue and white, are substituted for the red patch, the cell does not respond even though the dominant reflected wavelength is red. In such conditions, a human sees these latter patches as green, blue and white. This suggests that the wavelength-selective cells of V1 are concerned with the component wavelengths reflected from a surface, whereas the cells in V4 are concerned with the colour of a surface. This finding is consistent with lesion studies, which have shown that the removal or damage of V4 in monkeys leaves them able to discriminate wavelength, but impaired on colour constancy (e.g. Wild *et al.*, 1985).

A further piece of evidence comes from 'split brain' patients. The corpus callosum connects the two hemispheres and is essential for assigning the correct colour to a surface when one surface is presented in one hemi-field and the rest of the multicoloured display is presented to the other hemi-field (Land *et al.*, 1983). If the two are separated by 3.5 degree, the normal human

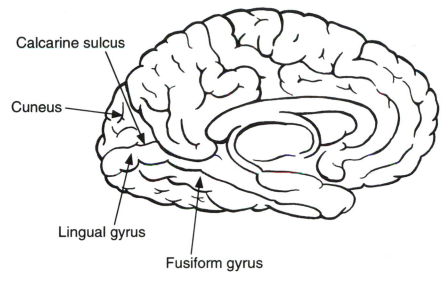

Calcarine sulcus

Cuneus

Lingual gyrus

Fusiform gyrus

Fig. 7.2. The positions of the lingual and fusiform gyri in the human cerebral cortex. (Redrawn from Zeki, 1993.)

brain can use the information coming from one hemi-field to generate the colour of the patch presented in the other hemi-field. If the corpus callosum is cut or absent, the colour-generating interactions cannot occur. Since V4 is the first visual area that has wavelength-selective cells and callosal connections extending beyond the representation of the 1 degree strip of the vertical meridian, it follows that V4 is the first possible area where such colour-generating interactions can occur.

Human area V4

The perception of colour in humans has been associated with activation of a ventromedial occipital area (in the collateral sulcus or lingual gyrus, see Fig. 7.2) in three separate PET studies (Corbetta *et al.*, 1991; Zeki *et al.*, 1991; Gulyas & Roland, 1991). It has been speculated that this area is the human homologue of V4 because monkey V4 contains colour-selective cells. The location of this area agrees well with the location of lesions associated with achromatopsia, which is close but medial to the posterior fusiform area activated by faces. That the colour- and face-selective areas are close to each other is consistent with evoked-potential studies from chronically implanted electrodes in epilepsy patients (Allison *et al.*, 1993; 1994). The proximity of these two areas would explain the frequent association of achromatopsia with pro-sopagnosia (the inability to recognise faces).

The neurons in monkey V4 are selective for features relevant to object recognition, including shape and colour (Zeki, 1983; Desimone & Schein, 1987), and, therefore, one would predict that the human homologue of V4 would be so also. However, of the two PET studies that examined colour and shape, one found that shape perception also activated the venteromedial occipitotemporal region (Corbetta *et al.*, 1991), but the other did not (Gulyas & Roland, 1991). Additional colour-selective regions have been reported in the lateral occipital cortex (Corbetta *et al.*, 1991). Moreover, lesions of monkey V4 produce significant impairments in form perception (Schiller & Lee, 1991), but form perception is usually spared in patients with achromatopsia. Also, the monkey V4 lesions do not seem to produce the profound and permanent colour impairment that is seen in patients with achromatopsia (Schiller & Lee, 1991; Heywood, Gadotti & Cowey, 1992). Therefore, although an area in human cerebral cortex has been located that is selective for colour, it may not be the homologue of monkey V4.

Key points

1. Surfaces and objects retain their colour in spite of wide ranging changes in the wavelength and energy composition of the light reflected from them. This is called colour constancy.

2. Edwin Land investigated colour constancy used a multicoloured display made of patches of paper of different colour pasted together (a Colour Mondrian).

3. When the spectral composition of the light illuminating the Mondrian was altered, the perceived colours of the patches remained the same. However, if a patch was viewed in isolation (the void viewing condition) the perceived colour of the patch corresponded to the wavelength composition of the light reflected from it. This suggests that the perceived colour of a patch is determined not only by the wavelength composition of the light reflected from it but also by the wavelength composition of the light reflected from the surrounding surfaces.

4. One physical constant of a surface that does not change with changes in the spectrum illumination is its reflectance. The biological correlate of reflectance is the perceived lightness of a surface.

5. The record of a scene in terms of areas that are lighter or darker is called its lightness record. Land's retinex theory proposes that in the visual

system the lightness records obtained simultaneously at three different wavelengths are compared to construct the colour of a surface.

6. Some neurons in monkey V1 and V2 are sensitive to the wavelength composition of light but do not show colour constancy. However, the responses of some cells in monkey V4 show the same colour constancy characteristics as a human observer viewing the same stimuli.

8

Object perception and recognition

From retinal image to cortical representation

In the primary stages of the primate visual system, such as Vl, objects are coded in terms of retinotopic co-ordinates, and lesions of Vl cause defects in retinal space, which move with eye movements, maintaining a constant retinal location. Several stages later in the visual system, at the IT, the receptive fields are relatively independent of retinal location, and neurons can be activated by a specific stimulus, such as a face, over a wide range of retinal locations. Deficits that result from lesions of IT are based on the co-ordinate system properties of the object, independent of retinal location. Therefore, at some point in the visual system, the pattern of excitation that reaches the eye must be transposed from a retinotopic co-ordinate system to a co-ordinate system centred on the object itself (Marr, 1982) (Table 8.1).

At the same time that co-ordinates become object centred, the system becomes independent of the precise metric regarding the object itself within its own co-ordinate system, that is to say that the system remains responsive to an object despite changes in its size, orientation, texture and completeness. Single-cell recording studies in the macaque suggest that for face processing these transformations occur in the anterior IT. The response of the majority of cells in the superior temporal sulcus (STS) is view selective and their outputs could be combined in a hierarchical manner to produce view-independent cells in the IT. As a result, selective deficits to higher visual areas, such as IT, cause the inability to recognise an object or classes of object. This defect in humans is called a visual *agnosia*.

Table 8.1. *A summary of Marr's model of object recognition. Marr viewed the problem of vision as a multi-stage process in which the pattern of light intensities signalled by the retina is processed to form a three-dimensional representation of the objects in one's surroundings*

Level	Description achieved
The raw primal sketch	Description of the edges and borders, including their location and orientation
The full primal sketch	Where larger structures, such as boundaries and regions, are represented
The 2½-dimensional sketch	A fuller representation of objects, but only in viewer-centred co-ordinates; this is achieved by an analysis of depth, motion and shading as well as the structures assembled in the primal sketch
The three-dimensional model	A representation centred upon the object rather than the viewer

Early visual processing

Visual recognition can be described as the matching of the retinal image of an object to a representation of the object stored in memory (Perrett & Oram, 1993). For this to happen, the pattern of different intensity points produced at the level of the retinal ganglion cells must be transformed into a three-dimensional representation of the object, which will enable it to be recognised from any viewing angle. The cortical processing of visual information begins in V1, where cells seem to be selective for the orientation of edges or boundaries. Boundaries can be defined not just by simple changes in luminance, but also by texture, colour and other changes that occur at the boundaries between objects. So what principles guide the visual system in the construction of the edges and boundaries that form the basis of the object representation?

The answer may lie, at least partially, with the traditional *gestalt* school of vision, which provides a set of rules for defining boundaries (see Table 8.2). For example, under the gestalt principle of *good continuity*, a boundary is seen as continuous if the elements from which it is composed can be linked by a

Table 8.2. *The gestalt principles of organisation*

Rule	Boundaries defined
Pragnanz	Every stimulus pattern is seen in such a way that the resulting structure is as simple as possible
Proximity	The tendency of objects near one another to be grouped together into a perceptual unit
Similarity	If several stimuli are presented together, there is a tendency to see the form in such a way that the similar items are grouped together
Closure	The tendency to unite contours that are very close to each other
Good continuation	Neighbouring elements are grouped together when they are potentially connected by straight or smoothly curving lines
Common fate	Elements that are moving in the same direction seem to be grouped together
Familiarity	Elements are more likely to form groups if the groups appear familiar or meaningful

straight or curved continuous line. Fig. 8.1*a* illustrates an illusory vertical contour that is formed by the terminations of the horizontal grating elements. There is no overall change in luminance between the left and right halves of the figure, yet a strong perceptual border exists. The operation of continuity can also be seen in Fig. 8.1*b*, where an illusionary bar seems to extend between the notches in the two dark disks. The illusory light bar is inferred by the visual system to join the upper and lower notches and the break in the central circle. In Fig. 8.1*c*, the illusory light bar is perceptually absent. Here the notches are closed by a thin boundary and each notch is, therefore, seen as a perceptual entity in its own right in accordance with the gestalt principle of closure. Psychologists have speculated that contours defined by good continuity were constructed centrally, rather than extracted automatically by neural feature detectors working at some stage of visual processing (Gregory, 1972). The illusory contours have, therefore, been given various labels including cognitive, subjective or anomalous. However, recent neurophysiological and behavioural results have disproved this idea and suggest that these illusory contours are extracted very early in the visual system.

Physiological studies have shown that specific populations of cells in early

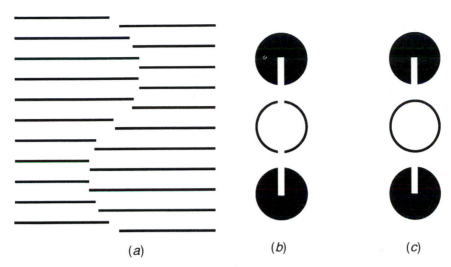

Fig. 8.1. Illusory contours. (a) Contour defined by the good continuation of line terminations of two gratings offset by half a cycle. (b) Illusory light bar induced by the good continuation of the edges of notches in the dark disks and gap in the central circle. (c) Illusory light bar of (b) disappears when the inducing notches are closed by a thin line (Redrawn from Peterhans & von der Heydt, 1991.)

visual areas (V1 and V2) do respond selectively to the orientation of contours defined by good continuity (Peterhans & von der Heydt, 1989; Grosof, Shapley & Hawken, 1993). Cells in V1 and V2 respond to illusory contours defined by the co-linearity of line terminations and signal the orientation of this illusory contour. Moreover, about one third of the cells tested in V2 responded to illusionary contours extending across gaps as well as they did to normal luminance contours, and the cells seem to exhibit equivalent orientation selectivity for real and illusory edges. This neurophysiological evidence is supported by the findings of Davis & Driver (1994), who used a visual search task to distinguish between early and late stages in the processing of visual information. For example, among many jumbled white letters, a single red one is instantly discerned (a phenomenon called 'pop out'), but a single L among many Ts needs more time to be detected. This result is taken to suggest that colour differences are extracted early in the visual system, but differentiation of similar letters is the result of more complex processing at a higher level. This procedure can be quantified by measuring the time it takes for a single odd feature to be detected among a number of background features. A rapid reaction time that is largely independent of the number of background features is taken to be indicative of processing at an early stage in the visual system. Davis and Driver used figures outlined by illusory contours based on

Fig. 8.2. The Kanizsa triangle.

the Kanizsa triangles (Fig. 8.2), and their results were consistent with the processing of these features occurring early in the visual system.

Therefore, the early cortical visual areas contain the neural machinery that is involved in the definition of boundaries in different regions of the retinal images. While many of these boundaries and contours are defined by luminance changes, analysis of subjective contours provides powerful supplementary cues to object boundaries (Perrett and Oram, 1993).

A visual alphabet?

As we move up the object-processing pathway in monkeys (V1–V2–V4–posterior IT–anterior IT) (see Fig. 8.3), the response properties of the neurons change. The receptive field of a cell gets significantly larger. For example, the average receptive field size in V4 is 4 degree^2, which increases to 16 degree^2 in the posterior IT and to 150 degree^2 in the anterior IT (Perrett and Oram, 1993). Most cells along the V4, posterior IT and anterior IT pathway also have

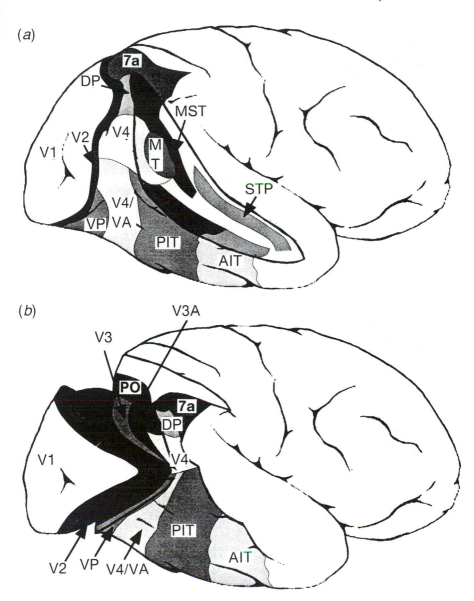

Fig. 8.3. The location of major visual areas in the macaque cerebral cortex. (a) The superior temporal sulcus has been unfolded so that the visual areas normally hidden from view can be seen. (b) The lunate, inferior occipital and parieto-occipital sulci have been partially unfolded. AIT, anterior inferior temporal cortex; DP, dorsal prelunate; MT, middle temporal also called V5; MST, medial superior temporal; PIT, posterior inferior temporal cortex; PO, parieto-occipital; STP, superior temporal polysensory; VA, ventral anterior; VP, ventral posterior. (Redrawn from Maunsell & Newsome, 1987.)

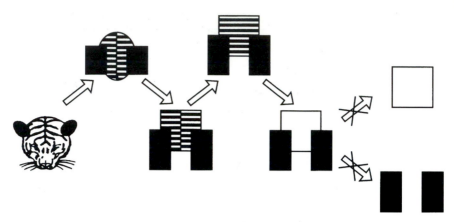

Fig. 8.4. An example of the procedures used by Tanaka and his colleagues in determining which features are critical for the activation of individual elaborate cells in IT. Among an initial set of three-dimension object stimuli, a dorsal view of the head of an imitation tiger was the most effective for the activation of a cell. The image was simplified while the responses of the cell were measured, the final result being that a combination of a pair of black triangles with a white square was sufficient to activate the cell. Further simplification of the stimulus abolished the responses of the cell. (Redrawn from Tanaka, 1992.)

receptive fields close to, or including, the fovea (75% of anterior IT cells included the fovea). The increase in receptive field allows the development of a visual response that is unaffected by the size and position of a stimulus within the visual field. The cells also respond to more complex stimuli. In V4 and in the posterior IT, the majority of cells have been found to be sensitive to the 'primary' qualities of a stimulus, such as colour, size or orientation, whereas cells in the anterior IT seem to be sensitive to complex shapes and patterns.

How cells in IT encode a representation of objects is a knotty problem. An interesting approach has been taken by Keji Tanaka. He has tried to determine the minimum features necessary to excite a cell in the anterior IT (Tanaka *et al.*, 1992; Tanaka, 1992). This method begins by presenting a large number of patterns or objects while recording from a neuron to find which objects excite that cell. Then the component features of the effective stimulus are segregated and presented singly or in combination (see Fig. 8.4), while assessing the strength of the cell's response for each of the simplified stimuli. The aim is to find the simplest combination of stimulus features to which the cell maximally responds. However, even the simplest 'real world' stimulus will possess a wide variety of elementary features, such as depth, colour, shape, orientation, curvature and texture and may show specular reflections and shading (Young, 1995). It is, therefore, not possible to present all the possible feature combina-

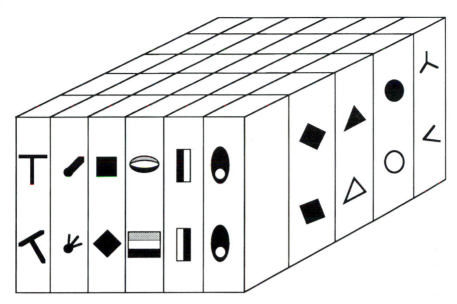

Fig. 8.5. Schematic diagram of the columnar organisation of the inferior temporal cortex. The average size of columns across the cortical surface is 0.5 mm. Cells in one column have similar but slightly different selectivities. (Redrawn from Tanaka, 1992.)

tions systematically, and the simplified stimuli that are actually presented in the cell's receptive field are typically a subset of the possible combinations. Hence, it is not possible to conclude that the best simplified stimulus is optimal for the cell, only that it was the best of those presented (Young, 1995).

Tanaka found a population of neurons in IT, called *elaborate cells*, that seemed to be responsive to simple shapes (Tanaka *et al.*, 1991; Fujita *et al.*, 1992). Cells in IT responsive to such simple stimuli seem be invariant with respect to the size and position of a stimulus and of the visual cues that define it (Sary, Vogels & Orban, 1993). Moreover, Tanaka found that closely adjacent cells usually responded to very similar feature configurations. In vertical penetrations through the cortex, he consistently recorded cells that responded to the same 'optimal' stimulus as for the first test cell tested, indicating that cells with similar preferences extend through most cortical layers. In tangential penetrations, cells with similar preferences were found in patches of approximately 0.5 mm². These results suggested to Tanaka that the cells in the IT are organised into functional columns or modules, each module specifying a different type of shape (Fig. 8.5). This hypothesis has been supported by an optical recording study that has shown a patchy distribution of staining on the surface of the IT, which would be consistent with a columnar organisation

(Wang, Tanaka & Tanifuji, 1994). If these modules are $0.5\,mm^2$ in width, then there could be up to 2000 within the IT. However, allowing for the fact that many may analyse the same type of shape, and many may analyse more complex patterns such as faces, the number of different simple shapes is probably only around 600 (Perrett & Oram, 1993).

This gave rise to the idea that these simple shapes form a *'visual alphabet'* from which a representation of an object can be constructed (Stryker, 1992). The number of these simple shapes is very small by comparison to the number of possible visual patterns, in the same way that the number of words that can be constructed from an alphabet is very large. Each cell would signal the presence of a particular simple shape if it was present in a complex pattern or object. Based on the responses of these elaborate cells, a representation could be derived in at least two ways. First, there could be a traditional hierarchy in which these elaborate cells would feed into a higher cell layer, whose cells respond preferentially to complex stimuli. The output of these cells would then signal the presence of a complex object to higher areas, such as the prefrontal cortex. Alternatively, there may be no upper layer. The pattern of responses across the various columns of elaborate cells may directly signal the presence of a complex object to a higher area, without having to converge on a cell in the IT sensitive to complex stimuli.

The shape selectivity of the elaborate cells is greater than that anticipated by many theories of shape recognition. For example, Irving Biederman (1987) described a theory of shape recognition that deconstructed complex objects into an arrangement of simple component shapes. Biederman's scheme envisaged a restricted set of basic three-dimensional shapes, such as wedges and cylinders, which he called *geons* (geometrical icons). Examples of these figures are shown in Fig. 8.6. These geons are defined only qualitatively. One example is thin at one end, fat in the middle and thin at the other. Such qualitative descriptions may be sufficient for distinguishing different classes of object, but they are insufficient for distinguishing within a class of objects possessing the same basic components (Perrett & Oram, 1993). Biederman's model is also inadequate for differentiating between perceptually dissimilar shapes (Fig. 8.7*b,c*) (Saund, 1992). Perceptually similar items (Fig. 8.7*a, b*) would be classified as dissimilar by Biederman's model. The single cell studies provide direct evidence that shape and curvature are coded within the nervous system more precisely than would be expected from Biederman's recognition by components model.

The concept of the visual alphabet assumes that an IT cell will reliably signal the presence of the particular simple shape that excites it regardless of whatever else is present in the visual field. However, as Malcolm Young has

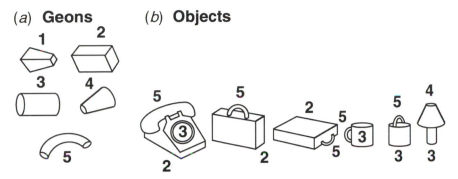

Fig. 8.6. (a) Examples of the simple, volumetric shapes (geons) proposed by Irving Biederman to form a basis of object perception. On the right side of the figure are examples of how these simple shapes could be used as building blocks to form complex objects. (Redrawn from Biederman, 1987.)

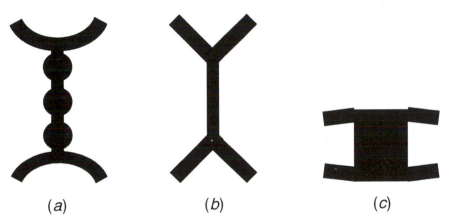

Fig. 8.7. Perceptual similarity of shapes; contrary to the predictions of the model of Biederman. The perceptual similarity of the shapes (a) and (b) appears greater than that between (b) and (c) (Redrawn from Saund, 1992; Perrett & Oram, 1993.)

pointed out, this may not always be the case (Young, 1995). In the example shown in Fig. 8.8a, (Cell 2, stimulus I), Tanaka's simplification procedure converged on an inverted T shape as the preferred simple shape for this cell. Any more complex object that contains this simple shape should evoke a strong response from the cell, as the cell is supposed to signal the presence of the shape. Just such an example is present by stimulus K to Cell 2 in Fig. 8.8, together with the cell's response to it. The cell did not respond well to a + shape, in which the preferred simple shape is still present, in concert with a bar below its centre. Therefore, the presence of other visual features can

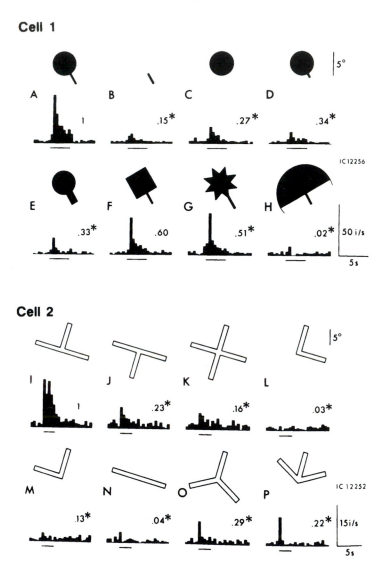

Fig. 8.8. An example of the responses of two separate cells to a number of stimuli used in Tanaka's simplification procedure. For Cell 1 the simplification procedure converged on a black circle and attached bar (stimulus A). For Cell 2 the simplification procedure converged on an inverted T (stimulus I). If the visual alphabet theory is correct, then a cell should respond strongly to any more complex object that contains the optimal stimulus shape as determined by the simplification procedure. For Cell 2, a more complex object containing this shape is stimulus K. The response of the cell to this stimulus is very weak, which contradicts the visual alphabet theory. (Reproduced with permission from Tanaka *et al.*, 1991. Copyright (1991) The American Physiological Society.)

disrupt the response of a cell to it's preferred shape, a result that is the opposite of that assumed in the visual alphabet conception of IT. The cell may be signalling a combination of the presence of its preferred shape and the absence of something else (Young, 1995). The characteristics of what must be absent for the cell to respond have not been defined, except to show that a bar below the centre of the inverted T is a member of this set, and so the simplification protocol has gone only half-way towards defining the conditions sufficient for the cell's response. If other cells behave in a similar way, the characterisation by this method of the simple shapes preferred by IT cells cannot be sufficient to account for the performance of the cells in the recognition of even slightly more complex objects.

Complex objects in three-dimensions: face cells

There is evidence that the cellular coding of at least some complex patterns and objects does not remain as a collection of separate codes for its component shapes. The most studied example is the *face cell*. Since the early 1970s it has been known that there are neurons in the monkey temporal visual cortex that are sensitive to faces, hands and other complex biological stimuli. Those responsive to faces are localised in the IT, and the banks and walls of the STS. The optimal stimuli of a large proportion of these cells cannot be deconstructed into simpler component shapes. In general, cells that are responsive to faces show virtually no response to any other stimulus tested (such as textures, gratings, bars and the edges of various colours) but respond strongly to a variety of faces, including real ones, plastic models and video display unit images of human and monkey faces. The responses of many face cells are size and position invariant; the cell's response is maintained when there is a 12-fold change in the size of the face, or if the position of the face within the cell's receptive field is altered (e.g. Rolls & Baylis, 1986; Tovée, Rolls & Azzopardi, 1994). Some face cells do not respond well to VDU images of faces that have had the components rearranged, even though all the components are still present and the outline is unchanged (e.g. Perrett, Rolls & Caan, 1982; Perrett *et al.*, 1992). Face cells are sensitive to the relative position of features within the face; particularly important is intereye distance, distance from eyes to mouth, and the amount and style of hair on the forehead (Yamane, Kaji & Kwano, 1988). Moreover, presentation of a single facial component elicits only a fraction of the response generated by the whole face, and removal of a single component of a face reduces, but does not eliminate, the response of a cell to a face. Face cells also continue to respond to VDU images of faces that have been low- or high-pass band filtered so that they have no spatial frequencies in common , that have had the colour removed or altered or that have had the contrast reduced to a very low level. Line drawings of faces produce a weak response. These complex neuronal properties suggest that this class of cell is truly face selective and not responding to some other characteristic of the visual image (for a review see Tovée & Cohen-Tovée, 1993).

Most face cells in the anterior IT and STS are selective for the viewing angle, such as the right profile of a face in preference to any other viewing angle. These cells are described as view-dependent or viewer-centred. A small proportion of the cells are responsive to an object irrespective of its viewing angle. These view-independent or object-centred cells may be formed by combining the outputs of several view-dependent cells. For example, view-independence could be produced by combining the responses of the view-dependent cells found in the STS. This hierarchical scheme would suggest that the response latency of such view-independent cells would be longer than that of the view-dependent cells, which proves to be the case. The mean latency of view-independent cells (130 ms) was significantly greater than that for view-dependent cells (119 ms) (Perrett *et al.*, 1992).

Single cell recording studies and optical recording techniques suggest that face cells, like elaborate cells, have a columnar organisation. Face cells responsive to a particular aspect of a face, such as the angle at which the head is observed, are found grouped together in columns that run perpendicular to the cortical surface (Harries & Perrett, 1991; Wang *et al.*, 1994).

The grandmother cell?

Temporal lobe face cells appear superficially to resemble the *gnostic units* proposed by Konorski (1967) or the *cardinal cells* proposed by Barlow (1972). These units were described as being at the top of a processing pyramid that began with line and edge detectors in the striate cortex and continued with detectors of increasing complexity until a unit was reached that represented one specific object or person, such as your grandmother, leading to the name by which this theory became derisively known. This idea had two serious problems. First, the number of objects you meet in the course of your lifetime is immense, much larger than the number of neurons available to encode them on a one to one basis. Second, such a method of encoding is extremely inefficient as it would mean that there would need to be a vast number of uncommitted cells kept in reserve to code for the new objects that one would be likely to meet in the future.

Although individual cells respond differently to different faces, there is no evidence for a face cell that responds exclusively to one individual face (Young & Yamane, 1992; Rolls & Tovée, 1995). Face cells seem to comprise a distributed network for the encoding of the faces, just as other cells in the IT probably comprise a distributed network for the coding of general object features. Faces are thus encoded by the combined activity of *populations* or *ensembles* of cells. The representation of a face would depend on the emergent spatial and temporal distribution of activity within the ensemble (Young & Yamane, 1992; Rolls & Tovée, 1995). Representation of specific faces or objects in a population code overcomes the two disadvantages of the grandmother cell concept. First, the number of faces encoded by a population of cells can be much larger than the number of cells that make up that population. So it is unnecessary to have a one to one relationship between stimulus and cell. Second, no large pool of uncommitted cells is necessary. Single-cell experiments have shown that the responses of individual neurons within a population alter to incorporate the representation of novel stimuli within the responses of existing populations (Rolls *et al.*, 1989; Rolls, Tovée & Ramachandran, 1993).

The size of the cell population encoding a face is dependent on the 'tuning' of individual cells. That is to say, how many or how few faces do they respond to? If they respond to a large number of faces, then the cell population of which they are a part must be large to signal accurately the presence of a particular face. A large cell population containing cells responsive to a large number of faces is termed *distributed* encoding. If a cell responds to only a small number of specific faces, then only a small number of cells in the population is necessary to distinguish a specific face. This is termed *sparse* encoding. Single-cell recording experiments in monkey IT have found that the face-selective neurons are quite tightly tuned and show characteristics consistent with sparse encoding (Young & Yamane, 1992; Rolls & Tovée, 1995). In a recent experiment, Edmund Rolls and myself recorded the responses of face cells to 68 visual stimuli (23 faces and 45 non-faces). Examples of these images are shown in Fig. 8.9. The responses of the neurons were tightly tuned to a subgroup of the faces shown, with very little response to the rest of the faces and the non-face stimuli (Fig. 8.10). These results suggest that the cell populations or ensembles may be as small as a hundred neurons.

Visual attention and working memory

Despite the vast number of neurons that compose the visual system, its ability to process fully and store in memory distinct, independent objects is strictly

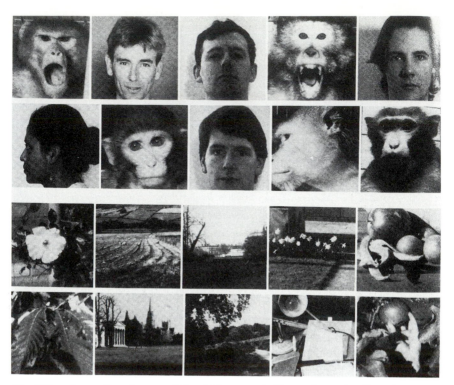

Fig. 8.9. Examples of some of the 23 faces and 45 non-face stimuli used to stimulate face cells. (Reproduced with permission from Rolls & Tovée, 1995. Copyright (1995) The American Physiological Society.)

limited. Robert Desimone has suggested that objects must compete for attention and processing 'space' in the visual system, and that this competition is influenced by both automatic and cognitive factors (Desimone *et al.*, 1995). The automatic factors are usually described at *pre-attentive* (or *bottom-up*) processes and the cognitive factors as *attentive* (or *top-down*) processes. Pre-attentive processes rely on the intrinsic properties of a stimulus in a scene, so that stimuli that tend to differ from their background will have a competitive advantage in engaging the visual systems attention and acquiring processing space. So, for example, a ripe red apple will stand out against the green leaves of the tree. The separation of a stimulus from the background is called *figure-ground segregation*. Attentive processes are shaped by the task being undertaken, and can override pre-attentive processes. So, for example, it is possible to ignore a red apple and concentrate on the surrounding leaves. This mechanism seems to function at the single cell level. When monkeys attend to a stimulus at one location and ignore a stimulus at another, microelectrode

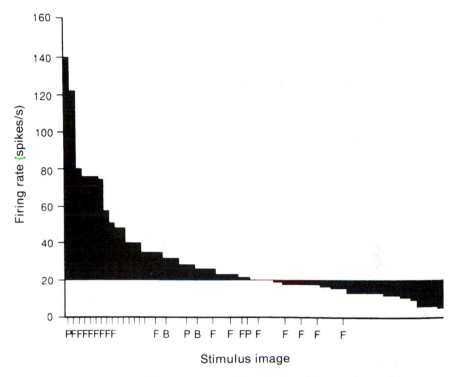

Fig. 8.10. Distribution of firing rates to all the 68 stimuli for one face cell in monkey IT. The firing rate of the neuron is shown on the ordinate, the spontaneous firing rate of the neuron was 20 spikes/s, and the histogram bars are shown as changes of firing rate from the spontaneous rate (i.e. neuronal responses) produced by each stimulus. Stimuli that are faces viewed frontally are marked F, those in profile are marked P. B refers to images of scenes that included a whole person, or other body parts such as hands or legs. The non-face stimuli are unlabelled. (Reproduced with permission from Rolls & Tovée, 1995. Copyright (1995) The American Physiological Society.)

recording shows that IT cell responses to the ignored stimulus are suppressed (Moran and Desimone, 1985). The cell's receptive field seems to shrink around the attended stimulus.

Analogous processes seem to occur within short-term visual memory. The effect of prior presentation of visual stimuli can be in either of two ways: suppression of neural response or by enhancement of neural response. Repeated presentation of a particular stimulus reduces the responses of IT neurons to it, but not to other stimuli. This selective suppression of neural responses to familiar stimuli may function as a way of making new or unexpected stimuli stand out in the visual field. This selective suppression can be found in monkeys passively viewing stimuli and even in anaesthetised animals (e.g. Miller, Gross & Gross, 1991), suggesting it is an automatic process that acts as a form of temporal figure-ground mechanism for novel stimuli and is independent of cognitive factors (Desimone *et al.*, 1995).

Enhancement of neural activity has been reported to occur when a monkey is actively carrying out a short-term memory task, such as *delayed matching to sample (DMS)*. In the basic form of this task, a sample stimulus is presented followed by a delay (the retention interval) and then by a test stimulus. The monkey has to indicate whether the test stimulus matches or differs from the sample stimulus. Some neurons in monkey IT maintain a high firing rate during the retention interval, as though they are actively maintaining a memory of the sample stimulus for comparison with the test stimulus (Miyashita & Chang, 1988). However, if a new stimulus is presented during the retention interval, the maintained neural activity is abolished (Baylis and Rolls, 1987). This neural activity seems to represent a form of visual rehearsal that can be easily disrupted (just as rehearsing a new telephone number can be easily disrupted by hearing new numbers), but this still may be an aid to short-term memory formation (Desimone *et al.*, 1995).

In another form of DMS task, a sample stimulus was presented followed by a sequence of test stimuli and the monkey had to indicate which of these matched the sample. Under these conditions, a proportion of IT neurons gave an enhanced response to the test stimulus that matched the sample stimulus (Miller & Desimone, 1994). Desimone has suggested that the basis of this enhanced response lies in signals coming in a top-down direction from the ventral prefrontal cortex, an area that has been implicated in short-term visual memory (Wilson *et al.*, 1993). Like IT neurons, some neurons in ventral prefrontal cortex show a maintained firing rate during the retention interval. This maintained firing is temporarily interrupted by additional stimuli shown during the retention interval, but the activity rapidly recovers. Desimone speculates that this maintained information about the sample stimulus may be fed back from the prefrontal cortex to the IT neurons so that they give an enhanced response to the correct test stimulus (Desimone *et al.*, 1995).

Pre-attentive memory processes are sensitive to stimulus repetition and automatically bias visual processes towards novel or infrequent stimuli (Fig. 8.11). Attentive processes are important when we search for a particular stimulus in a temporal sequence of different stimuli. Together these two types of process determine which stimulus in a crowded scene will capture our attention.

Top down (active)
working memory

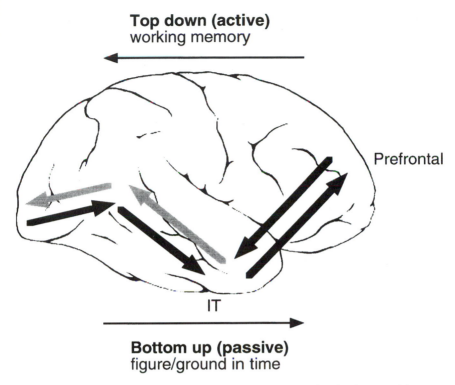

Prefrontal

IT

Bottom up (passive)
figure/ground in time

Fig. 8.11. Dual mechanisms of short-term memory. Simple stimulus repetition engages passive, or bottom-up, mechanisms in IT cortex and possibly earlier visual areas. These mechanisms mediate a type of memory that assists the detection of novel or not recently seen stimuli, like a form of temporal figure–ground segregation. By contrast, working memory is believed to involve an active, or top-down mechanism, in which neurons in IT cortex are primed to respond to specific items held in short-term memory. This priming of IT neurons seems to require feedback from the prefrontal cortex. (Redrawn from Desimone *et al.*, 1995.)

Visual imagery and long-term visual memory

Visual areas in the brain may also have a role to play in long-term memory and visual imagery. If we close our eyes and summon up the image of a particular person, object or scene, it seems that at least some of our visual areas become active. Although long-term memory is thought to be mediated primarily by the hippocampus and its associated areas, these areas all have extensive back projections, both directly and indirectly, to the visual system. Functional imaging studies (such as PET and fMRI) have shown that in recall of objects the higher visual areas are active and that damage to these areas impairs recall (Roland & Gulyas, 1994; Kosslyn & Oschner, 1994; Le Bihan *et al.*, 1993). However, there has been considerable debate about the extent of the reactivation of the visual system and whether it involves the early visual areas, such as V1 and V2. Kosslyn & Oschner (1994) have argued that mental imagery requires the activation of all the cortical visual areas to generate an image, whereas Roland & Gulyas (1994) have pointed out that if the brain has already produced a representation of a particular stimulus in the temporal or parietal cortex, why should it need to do it all over again? The evidence for either argument is inconclusive. Using PET, Roland has reported that early visual areas do not become active (Roland & Gulyas, 1994), but most other PET studies and fMRI studies have shown activation of these areas (Kosslyn & Oschner, 1994; Le Bihan *et al.*, 1993). Studies from brain-damaged subjects are equally contradictory (see Roland & Gulyas, 1994; Kosslyn & Oschner, 1994; Moscovitch, Behrmann & Winocur, 1994). However, at present the weight of evidence from both functional imaging and clinical studies suggests that all the cortical visual areas are active during visual imagery and recall from long-term visual memory.

Key points

1. The pattern of different luminance intensity points produced at the level of the retinal ganglion cells must be transformed into a three dimensional representation of the object, which will enable it to be recognised from any viewing angle.
2. Some aspects of the traditional gestalt school of perception may guide the visual system in the construction of the edges and boundaries that form the basis of the object representation. However, these seem to be automatic, rather than cognitive processes, and are implemented in early visual areas (such as in V1 and V2).
3. The response properties of visual neurons become more complex as one moves up the visual system, and neurons in monkey IT, called elaborate cells, seem to be responsive to simple shapes. The elaborate cells seem to be organised into functional columns or modules, each module specifying a different type of shape.
4. It has been suggested that the simple shapes coded for by the elaborate cells can form a 'visual alphabet' from which a representation of an object can be constructed
5. Some neurons seem to be responsive to more complex shapes than are the elaborate cells; some of these neurons are the face cells, which may

represent the neural substrate of face processing. These neurons also seemed to have a columnar organisation.

6. Neurons seem to comprise a distributed network for the encoding of stimuli, just as other cells in the IT probably comprise a distributed network for the coding of general object features. Stimuli are, thus, encoded by the combined activity of populations or ensembles of cells.

7. The activity of visual neurons in the monkey IT seems to be important in the maintenance of short-term visual memory. This activity is at least partially dependent on feedback projections from areas in the frontal cortex, which have been implicated in visual working memory.

8. In visual imagery, when we close our eyes and summon up the image of a particular person, object or scene, it seems that at least the higher visual areas become active. This activation is believed to be mediated by feedback projections from higher areas, such as the hippocampus.

9

Face recognition and interpretation

What are faces for?

The recognition and interpretation of faces and facially conveyed information is a complex, multi-stage process. A face is capable of signalling a wide range of information. It not only identifies the individual but also provides information about a person's gender, age, health, mood, feelings, intentions and attentiveness. This information, together with eye contact, facial expression and gestures, is important in the regulation of social interactions. It seems that the recognition of faces and facially conveyed information is separate from the interpretation of this information.

Face identification

The accurate localisation in humans of the area, or areas, important in the recognition of faces and how it is organised has plagued psychologists and neuroscientists for some years. The loss of the ability to recognise faces (*prosopagnosia*) has been reported in subjects with damage in the region of the occipito-temporal cortex, but the damage, whether through stroke or head injury, is usually diffuse. The subjects suffer not only from prosopagnosia but usually from other forms of agnosias too, and often impaired colour perception (achromatopsia). However, PET scanning has allowed more accurate localisation (see Fig. 9.1), and these studies have suggested that the posterior *fusiform gyrus* is activated in tasks requiring identification based on general facial features, such as face matching or gender discrimination (Haxby *et al.*, 1991;

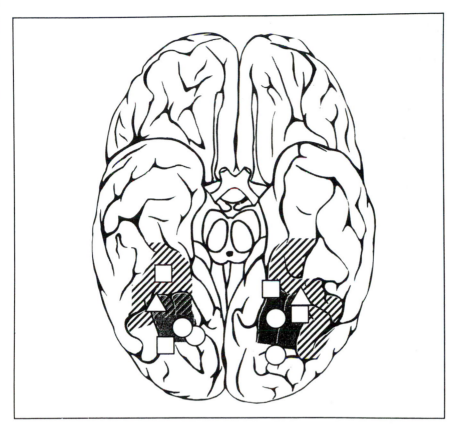

Fig. 9.1. The underside of the human cerebral cortices. For the visual-evoked potential studies, the stippled regions indicate where N200 (face-specific) potentials were recorded and the cross hatching from where colour potentials were recorded. For the PET studies, circles indicate centres of colour activation, triangles indicate centres of activation for a face-matching task and squares indicate centres of activation in a face-identification task. (Reproduced with permission from Tovée, 1995b. Copyright (1995) Current Biology.)

Sergent, Ohta & MacDonald, 1992), whereas identification of a unique individual's face activates the mid-fusiform gyrus (Sergent *et al.*, 1992). More anterior regions seem to be involved in activities such as short-term visual memory (Sergent *et al.*, 1992).

This is consistent with a study by Truett Allison and his colleagues, which recorded field potentials from strips of stainless steel electrodes resting on the surface of the extrastriate cortex in epileptic patients being evaluated for surgery. The electrodes were MR imaged to allow precise localisation in relation to the sulci and gyri of the occipito-temporal cortex. They recorded a large amplitude negative potential (N200) generated by faces and not by the other categories of stimuli they used (Allison *et al.*, 1994). This potential was generated bilaterally in regions of the mid-fusiform and inferior temporal gyri. Interestingly, electrical stimulation of this area caused transient prosopagnosia. To confirm this result, Allison then used fMRI techniques to study blood flow during the same face recognition task and found activation of the same areas of the brain as indicated by field potential recording (Puce *et al.*, 1995). This distribution of response properties within the fusiform gyrus suggests that in humans, as in monkeys, there is an increasing complexity of representation as one moves from posterior to anterior fusiform gyrus, with a build-up of general representations of faces to produce a representation of a unique face.

Prosopagnosia is frequently associated with achromatopsia, suggesting that the areas mediating these functions are in close proximity. Allison and his colleagues recorded potentials evoked by red and blue coloured checkerboards (Allison *et al.*, 1993). These potentials were localised to the posterior portion of the fusiform gyrus and extended into the lateral portion of the lingual gyrus. Electrical stimulation of this area caused significant colour effects in the patient's visual perception, such as coloured phosphenes and, less commonly, colour desaturation (Allison *et al.*, 1993). This finding is consistent with the position of lesions causing achromatopsia (Zeki, 1990), post-mortem anatomical studies of the human cortex (Clarke, 1994) and PET scan studies (Corbetta *et al.*, 1991; Watson, Frackowiak & Zeki, 1993a); this region may be the human homologue of monkey V4.

Laterality and face recognition

There is considerable evidence from psychophysical experiments and brain-damaged subjects that the left and right hemispheres process face information differently and that right hemisphere damage may be sufficient to cause prosopagnosia. Presentation of faces to the left visual field (and, therefore, initially to the right hemisphere) of normal subjects leads to faster recognition than presentation to the right visual field (left hemisphere) and to greater accuracy in recognition. The right-hemisphere advantage disappears when faces are presented upside down, and right-side damage disrupts recognition of upright faces but not inverted faces (Yin, 1969; 1970). It seems that, in the right hemisphere, upright faces are processed in terms of their feature configuration, whereas inverted faces are processed in a piece-meal manner, feature by feature (Carey & Diamond, 1977; Yin, 1970). In the left hemisphere, both upright and inverted faces seem to be processed in a piecemeal manner (Carey & Diamond, 1977). Interestingly, Allison and his colleagues reported that normal and inverted faces produce the same N200 pattern in the left hemisphere, but in the right hemisphere the N200 potential was delayed and much smaller in amplitude in response to the inverted face.

These findings are consistent with the clinical and neuropsychological studies that suggest that patients with brain damage in the right hemisphere show a greater impairment in face-processing tasks than patients with the equivalent damage in the left hemisphere (De Renzi et al., 1994). Although the complete loss of face recognition capacities seems to be associated with bilateral damage (Damasio, Tranel & Damasio, 1990), there are suggestions that unilateral right-hemisphere damage might be sufficient (De Renzi et al., 1994; Sergent & Signoret, 1992). One of the most common causes of prosopagnosia is cerebrovascular disease. The infero-medial part of the occipito-temporal cortex (including the fusiform gyrus, lingual gyrus and the posterior part of the parahippocampal gyrus) is supplied by branches of the posterior cerebral arteries, which originate from a common trunk, the basilar artery. It is, therefore, common to find bilateral lesions when the basilar artery is affected. Moreover, when a unilateral posterior cerebral artery stroke does occur, it is common for further ischaemic attacks to occur in the cortical area served by the other posterior cerebral artery (Grusser & Landis, 1991). It is, therefore, not surprising that prosopagnosic patients are commonly found with bilateral lesions of the occipito-temporal cortex. However, Landis, and his colleagues (1988) report the case of a patient who had become prosopagnosic after a right posterior artery stroke, and who died 10 days later from a pulmonary embolism. The autopsy revealed a recent, large inferomedial lesion in the right hemisphere and two older clinically silent lesions, a microinfarct in the lateral left occipito-parietal area and a right frontal infarct. The short delay between symptom and autopsy suggests that a right medial posterior lesion is sufficient for at least transient prosopagnosia. Although it might be argued that in this case some recovery of face processing ability might have occurred with time, there is also evidence of unilateral right-hemisphere damage producing long-lasting prosopagnosia. Grusser and Landis (1991) cite more than 20 prosopagnosic patients who are believed to have unilateral, right-hemisphere brain damage on the basis of intraoperative and/or neuroimaging techniques. In many of these patients prosopagnosia has existed for years. Although intraoperative findings and neuroimaging techniques are less precise than autopsy results, and small lesions may go undetected in the left hemisphere, the lesion data in humans do suggest that face processing is primarily, if not exclusively, a right-hemisphere task.

Face interpretation and the amygdala

Although recognition of faces or facial expression seems to occur in the temporal visual cortex, the actual interpretation of facially conveyed information seems to occur in later structures, such as the *amygdala*. The amygdala (so called for its resemblance to an almond in its size and shape) receives inputs from the association areas of three sensory modalities (visual, auditory and somatosensory) and from polysensory areas, such as the dorsal bank of the superior temporal sulcus (STS) (Amaral *et al.*, 1992). The amygdala projects directly to the striatum, the hypothalamus and the brain-stem centres. So it is directly linked with uni- and polymodal sensory regions on the input side and with motor, endocrine and autonomic effector systems on the output side. In monkeys, bilateral removal of the amygdala produces a permanent disruption of social and emotional behaviour (part of the *Kluver–Bucy syndrome*). This evidence suggests that the amygdala is an important route through which external stimuli could influence and activate emotions. This hypothesis is supported by models of the functional connectivity of the primate cortex, which show the amygdala to be a focal point in the passage of sensory information to the effector areas (Young & Scannell, 1993). However, this simple answer to the role of the amygdala is complicated by the extensive backprojections to the cortex, which not only reciprocate the afferent connections but also reach many other regions of the association cortex. This is most marked in the visual system, where amygdala efferents terminate in every visual region of the temporal and occipital cortex and may be involved in the modulation of sensory processing by affective states (Fig. 9.2).

As mentioned, neurons in the monkey STS (an area which projects strongly to the amygdala) are sensitive to the facial expression, direction of gaze and orientation of faces (Hasselmo, Rolls & Baylis, 1989; Perrett *et al.*, 1992); neurons in the amygdala also show selectivity to faces and features such as the direction of gaze (Brothers & Ring, 1993). One of the functions of the amygdala is to link the perception of a facial expression with a particular emotion in the observer. For example, an expression of fear may induce a shared feeling of fear in the observer. The amygdala helps give 'meaning' to an expression and aids in its interpretation. The amygdala also plays a role in 'fine tuning' our discrimination of facial expressions. In humans, the location of the amygdala, buried deep in the temporal lobe, means that selective damage to the amygdala is very rare. However, an example of this condition has recently been reported by Damasio and his colleagues. They have studied a 30-year-old woman (S. M.), of normal intelligence, who suffers from Urbach–Wiethe disease. This is a rare congenital condition that leads in

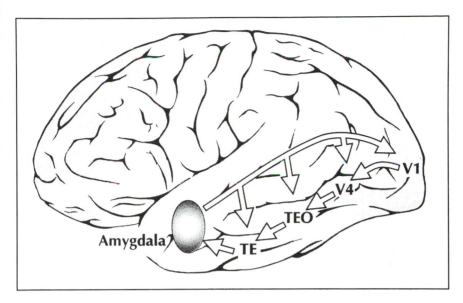

Fig. 9.2. Schematic illustration showing the relationship of the primate amygdala with the ventral stream of the visual system. Visual information is processed in hierarchical fashion from V1 to IT. The amygdala receives a substantial input from the anterior IT (labelled as TE), and projections from the amygdala pass back to all visual areas. (Reproduced with permission from Tovée, 1995b. Copyright (1995) Current Biology.).

around 50% of cases to the deposition of calcium in the amygdala during development. In the case of S. M., CT and MRI scans have shown that this condition has caused a nearly complete bilateral destruction of the amygdala, while sparing the hippocampus and other neocortical structures (see Figs. 1.3 and 1.4, p. 6 and 7) (Tranel & Hyman, 1990; Nahm *et al.*, 1993). S. M.'s face recognition capabilities seem to be normal. She could recognise familiar faces and learn to recognise new faces (Adolphs *et al.*, 1994). However, when tested with faces showing six basic emotions (happiness, surprise, fear, anger, disgust and sadness) and asked to rate the strength of those emotions, she displayed a severe impairment in rating the intensity of fear relative to the ratings of normal subjects and brain-damaged controls.

S. M. was then asked to rate the perceived similarity of different facial expressions (Adolphs *et al.*, 1994). The results from normal subjects suggested that facial expressions have graded membership in categories of emotion and that an expression can be a member of more than one emotion category. For example, happy and surprised expressions were rated as very similar, and elements of one expression may be present in the other. The results from S. M. did not show this graded categorisation, suggesting that S. M. cannot interpret a blend of emotions expressed by a face. Instead,she categorises the expression on the basis of the prototypical emotion expressed.

The expression recognition impairment found in S. M. seems to be a developmental effect. Two adult men, who survived herpes simplex encephalitis with complete bilateral lesions of the amygdala and additional temporal lobe structures, were tested on the same set of experiments as S. M. and were found to be unimpaired in recognising facial expressions of emotion, including fear (Hamann *et al.*, 1996). The main difference between these two men and S. M. seems to be the timing of the amygdala damage. The two men had their lesions late in life (after 50 years of age), whereas S. M., suffering from a congenital condition, was lesioned early in life. To produce a significant impairment of the recognition of facial expression, the damage to the amygdala must occur during development. The linkage of an expression to an emotion in ourselves helps us learn to discriminate between, and grade the intensity of, facial expressions. However, once we have learnt the meaning of an expression, we need not feel the emotion normally linked to an expression to recognise it. Therrefore, damage to the adult amygdala may not impair our recognition of an expression, but it does seem to impair our emotional reaction to it.

The frontal cortex and social interaction

The frontal cortex also plays an important role in the interpretation of visual information, such as that conveyed by faces. It has been known for many years that damage to the frontal lobes leads to a disruption of emotional and social behaviour. A particularly well-known example is the case of Phineas P. Gage. Phineas was a 25-year-old construction foreman for the Rutland and Burlington Railroad Company in New England. In 1843, in order to lay new railway tracks across Vermont, the railroad company carried out controlled blasting to level the uneven ground. Phineas was in charge of this blasting. His job was to drill holes in the rock, partially fill them with explosive powder, cover the powder with sand and then tamp it down with an iron rod before laying a fuse. On the 13th of September, Phineas made a mistake. In a moment of abstraction, he began tamping directly over the explosive powder, before his assistant had covered it with sand. The resultant explosion threw the fine-pointed, 3 cm thick, 109 cm long tamping rod up into the air like a javelin, and it landed several yards away. Unfortunately, Phineas had been bending over the tamping rod and during its trajectory it passed through his face, skull and brain (Fig. 9.3). Although, he was initially stunned, he rapidly regained consciousness and was able to talk and even walk away from the site of his injury, if a little unsteadily. However, he was subsequently a changed man. Prior to his injury he had been an intelligent, responsible, socially well-adapted individual. Subsequent to his injury, although his intelligence and other faculties were unimpaired, he became unreliable and offensive, with no respect for social conventions. He lost his job and led a wandering, transient existence until his death 12 years later. No autopsy was performed, but his skull (and the fateful tamping rod) were recovered for science and have been on display at the Warren Anatomical Medical Museum at Harvard University.

Recently, a team led by Damasio has tried to calculate the brain areas damaged in the accident, based on the skull and the contemporary account of the attending physician, Dr John Harlow (Damasio *et al.*, 1994). They conclude that damage was limited to the ventromedial region of both frontal lobes, while the dorsolateral region was spared (Fig. 9.3). This pattern of damage has been found in modern patients who have an inability to make rational decisions in personal and social matters and are impaired in processing emotion. Damasio has suggested that emotion, and its underlying neural substrate, participate in decision making within the social domain, and that an important site for this process is the ventromedial frontal region. This region has reciprocal connections to subcortical structures, such as the amygdala and hypothalamus, which control basic biological regulation, the processing of information on emotion, and social cognition. The dorsolateral frontal region, by comparison, seems to be important for cognition concerning extrapersonal space, objects, language and arithmetic (Posner & Petersen, 1990).

Faces as a social semaphore

The primate face has undergone a remarkable transformation. Its neural innervation, musculature and flexibility have extensively increased, from the almost rigid mask of some New World monkeys to the flexible, highly mobile face of the great apes; it reaches its height of sophistication and elaboration in humans. What is the point of it all? It is not simply for identification purposes; a pattern of facial features unique to an individual need not be mobile to signal his or her identity. It seems that as primates have developed more complex social groups the primate face has developed into a kind of semaphore system, capable of signalling a wide range of complex social information. The recognition and interpretation of these social cues are extremely important for the smooth functioning of a social group and for an individual's place within the hierarchy of this group. The increasing complexity of facial musculature and innervation seems to have been paralleled by an

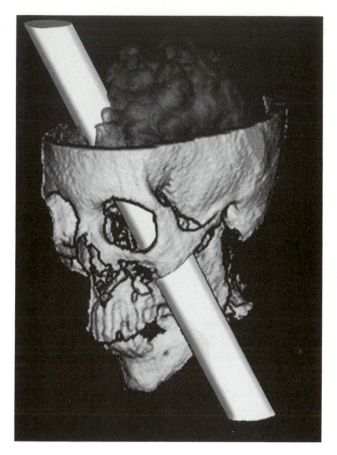

Fig. 9.3. View of the proposed position of the tamping rod in the skull and brain of Phineas Gage. (Redrawn from Damasio *et al.*, 1994.)

increasing sophistication of the neural representation of facially signalled information. However, there is strong evidence for a dissociation between recognition of facially conveyed information, such as identity and emotion, and the interpretation of this information. In both humans and monkeys, the recognition component seems to occur in the temporal visual cortex, whereas interpretation seems to occur in subsequent structures, such as the amygdala. The amygdala seems to be important both to interpret and to give meaning to the basic emotion of fear in facial expressions, and it seems to facilitate the differentiation of the blends of multiple emotions that the human face can signal. The development of elaborate and complex neural mechanisms for the recognition and interpretation of facially transmitted information has occurred; subjects with damage to this system, such as S. M., are impaired in social interactions and decision making (Tranel & Hyman, 1990; Nahm *et al.*, 1993).

Key points

1. Damage to certain areas of the brain can cause a specific form of agnosia called prosopagnosia. This is the inability to properly process visual information on facial identity or facially conveyed information, such as expression or direction of gaze.
2. In humans, the area of the brain that mediates face recognition and interpretation is the fusiform gyrus and, possibly, the lower part of the inferior temporal gyrus.
3. Face processing seems to show lateralisation. The right side of the human cortex seems to be specialised to process faces as a single pattern. The left side processes faces in a 'piecemeal' manner, that is feature by feature rather than as a single pattern.
4. Although recognition of faces or facial expression seems to occur in the fusiform gyrus and related areas, the actual interpretation of facially conveyed information seems to occur in later structures, such as the amygdala. Bilateral removal of the amygdala produces a permanent disruption of social and emotional behaviour (part of the Kluver–Bucy syndrome).
5. Another important area for the interpretation of emotion in the context of social interactions is the ventromedial frontal cortex. Damage to this area, such as in the famous case of Phineas Gage, disrupts the interpretation of information relating to emotion and a subject's ability to function socially.

10

Motion perception

The illusion of continuity

In determining the nature of the movement of an object or scene across the retina, the visual system has to determine whether the eyes are moving, the head or body is moving, or the object itself is moving. To determine whether the eyes are moving, it seems that the cortical motor areas that control eye movement simultaneously send a signal to the visual system (the *corollary discharge theory*). For example, if the eye muscles of volunteers are temporarily paralysed and they are asked to try and move their eyes, the volunteers report that the scene seems to jump to a new position even though their eyes do not move and the scene does not change (Stevens, *et al.*, 1976; Matin *et al.*, 1982).

It is important for the visual system to know about eye movements and to be able to compensate for their effects, as under normal circumstances our eyes are constantly moving. The reason for this constant movement can be found in the organisation of the retina. High-acuity colour vision is limited to the central 2 degrees of the visual field subserved by the fovea. Outside this small window, the spatial sampling of the retinal image declines sharply with increasing distance from the fovea (Perry & Cowey, 1985). Similarly, the packing of colour-sensitive cones declines by a factor of about 30 as one moves from central vision to 10 degrees of eccentricity (Curcio *et al.*, 1991). Indeed, as you read this page, only about 16 letters are fully processed, and the rest of the text can be turned into Xs without it being noticed or reading performance being impaired (Underwood & McConkie, 1985). This concentration on the central visual field is continued in the cortex. For example, in V1, three to six times as much space is devoted to the representation of central vision as is devoted to the periphery (Azzopardi & Cowey, 1993).

Fig. 10.1.   On the left is a photograph of Malcolm Young and his students in the laboratory. The visual system provides us with the sensation that all aspects of the scene are simultaneously processed. On the right is a processed version of the same photograph, giving an impression of what the retina of a viewer might signal to the brain about this scene. Only a restricted region corresponding to the central few degrees of vision is fully processed, our impression of a complete scene is built up from a number of 'snap shots' made by the constantly moving eye. (Reproduced with permission from Young, 1993b. Copyright (1993) Current Biology.)

So how, from this tiny visual window (see Fig. 10.1), is the high acuity, colour image of the world we believe we see constructed? It seems that despite our impression of a stable visual image, our eyes are always moving even when we are looking at a single object in a scene. This allows all or most of the features of a scene to be brought into the high-acuity centre of the visual field. Our visual image seems to be constructed by repeated foveation of different objects, or parts of objects, the use of short-term working memory of each snap shot and the predictive properties of the visual system (such as in 'filling in') (Young, 1993b). Our perception of the world is merely a 'best guess' of what is really out there, generated by our cortex.

Saccades

There are two forms of involuntary eye movement; these are called *micro saccades* and *saccades*. When we fixate a scene, our eyes are not absolutely still but make constant tiny movements (called microsaccades or tremors). These occur several times per second, are random in direction and are about 1–2 min of arc in amplitude. If an image is artificially stabilised on the retina, eliminating any movement relative to the retina, vision fades away after about a second and the image disappears. Artificially moving the image on the retina causes the perception of the image to reappear. The neurons of the visual system rapidly adapt to a stationary stimulus and become insensitive to its continued presence. Therefore, microsaccades are necessary to allow the perception of stationary objects.

When we visually explore our environment, our eyes do not move in smooth continuous movements. Instead, our eyes fixate an object for a brief period (around 500 ms) before jumping to a new position in the visual field (see Fig. 10.2). These rapid eye movements are called saccades. Saccades can reach very high velocities, approaching 800 degrees/s at their maximum. The size of the saccade is typically 12–15 degrees, but with significant numbers of both larger and smaller amplitudes.

Although a saccade may be used to foveate a moving stimulus, the eye must somehow subsequently track the stimulus as it moves through the visual field. This is done using *pursuit eye movements*, sometimes called *smooth eye movements*. Unlike the jerky saccades, pursuit movements are smooth. They are not ballistic. The neural signals being sent to the extraocular muscles, which mediate pursuit movement, are being constantly updated and revised, allowing the speed and direction of the pursuit movements to alter with changes in the speed and direction of the target. These movements have a maximum target velocity of around 30 degrees/s. When the whole visual scene moves, then a characteristic pattern of eye movements occurs, called an *opto-kinetic nystagmus* (*OKN*). An example of this situation is when you look out of the window of a moving vehicle, and for this reason OKN was once known as railway nystagmus. OKN has two components, called the fast and slow phases. In the slow phase, there is a smooth pursuit of the moving field, which stabilises the image on the retina. If the velocity of field movement increases above 30 degrees/s, the eyes lag progressively behind, and the stabilisation is less effective. The slow-pursuit phase alternates with fast, saccadic eye movements that return the eyes to the straight ahead position. The OKN seems to be a primitive form of eye movement control, designed to prevent displacement of the retinal image during locomotion.

Suppression of perception during saccades

During rapid eye movements, like those made while reading this book, you will not be conscious of a visual 'smear' caused by the movement of the image across the retina. During each saccad the retinal image is displaced at a speed of several hundred degrees/s, and such displacements of the image are per-

Fig. 10.2. The Russian experimenter Yarbus recorded the eye movements of subjects as they explored various images, such as woods or a female face. The stopping points of a subjects gaze are shown as points on the figures, which are joined by lines indicating the eyes' movement during a saccade. (Reproduced with permission from Yarbus, 1967. Copyright (1967) Plenum Publishing Corp.)

ceived as movement if they occur when the eyes are stationary. So why is our vision not always being interrupted by the smearing effects of the constant saccades our eyes are making, even when we observe a single stationary object? The obvious answer is that some form of suppression of the signal from the eye occurs when it makes a saccade. One can easily see this effect by looking in a mirror and changing the fixation from the image of the pupil to the edge of the eye. The movement your eye makes is invisible in the mirror. This is not because the movements are too small or too fast to be seen, as they are easily observed when looking at another person's eyes (Morgan, 1994). However, this is not to say that perception is entirely suppressed during a saccade. If you look at a rail track from a fast-moving train, the sleepers only become visible when you make a saccade against the direction of the train and, thus, briefly stabilise the image of the track on the retina.

This suppression of perception seems to be confined to the M pathway (Burr, Morrone & Ross, 1994). As the M pathway is primarily sensitive to motion and the P pathway is primarily sensitive to colour and high acuity, it is possible to tease out the influence on visual perception of the two systems by using the right stimuli. If one uses stimulus gratings of high spatial frequency, and which are equiluminant but differ in colour, then their perception is unaffected by saccades (Burr *et al.*, 1994). But if one uses coarse gratings containing no colour, then their perception is suppressed by saccades. Moreover, if the spectral sensitivity of a subject is measured for brief stimulus presentations either during or before a saccade, then the spectral sensitivity function during a saccade shows the sensitivity to wavelength expected of the P pathway, and the spectral sensitivity function outside the saccade is that expected of the M pathway (Uchikawa & Sato, 1995). Therefore, it seems that the M pathway is suppressed during saccades, but the P pathway is not.

The image during a saccade is moving very rapidly and would probably primarily stimulate the M pathway. It is, therefore, important to suppress its action to allow the creation of a stable image of the external world. The P pathway is probably not stimulated very much by the moving image during the saccade, so there has been no selective pressure to develop a mechanism to suppress its activity.

The neural basis of motion detection

The perception of movement of an external stimulus can be produced in a number of different ways (see Table 10.1). Movement perception seems to be mediated by the M pathway. The M pathway projects to areas V3 and V5 (otherwise called middle temporal area or MT), both directly and through the thick stripes of V2. Areas V3 and V5 have considerable interconnections and both project to the parietal cortex, where a spatial representation of the environment seems to be encoded.

Most cells in V3 are orientation selective and are believed to be concerned with processing dynamic form and *three-dimensional structure from motion (3D-*

Table 10.1. *Stimuli that cause the perception of motion*

Movement	Conditions
Real	An object is continuously displaced from one point to another
Apparent or stroboscopic	This is an illusion of movement that can be created by flashing one light on and then off, followed by flashing another light, displaced to one side, on and off 60 ms later
Induced	If the background surrounding an object moves one way, the object may be seen to move in the opposite way
Autokinetic	In a darkened room, a stationary light without spatial cues to place it in a certain position is seen to move erratically
Aftereffects	If an observer views a pattern that is moving in one direction and then views a stationary object, the object will appear to move in the opposite direction to that moved by the pattern. An example of this is the waterfall illusion

SFM) (Zeki, 1993). An example of 3D-SFM can be easily demonstrated. If a piece of wire is bent into a complex, three-dimensional shape and then illuminated such that it casts a shadow on a screen, an observer will not be able to determine the wire's shape from the shadow. However, if the wire is rotated, its three-dimensional shape is immediately apparent (Wallach & O'Connell, 1953). Another example is the perception of optic flow and motion in depth. Optic flow fields are very popular visual stimuli, and on a VDU screen they simulate the forward motion of an observer over a flat plain covered in bright dots of a uniform density (Fig. 10.3). PET scan studies have shown that the human equivalent of V3 is differentially activated by these optic flow fields (de Jong *et al.*, 1994).

V5 is an important area in the processing of visual information. It is perhaps analogous to area V4 in the P pathway, and a considerable amount of research has been focused on its function and organisation. In monkeys, lesions of V5 cause deficits in discriminating the direction of motion, and single-cell recording techniques have shown that all V5 neurons responded better to moving stimuli than to stationary stimuli and that most of them give the same response regardless of the colour or shape of the test stimulus (Albright, 1984). Each V5 neuron responds preferentially to a particular speed and direction of motion. Some of these neurons show a more complex analysis of motion. If a stimulus is made up of two gratings drifting across a VDU

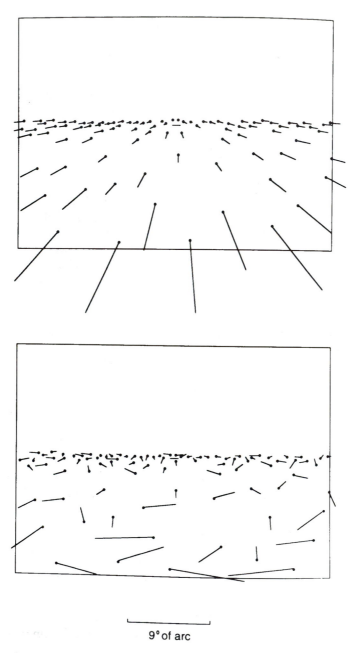

Fig. 10.3. Schematic diagram of two motion stimuli used to study the perception of motion in depth. The stimuli are moving dots, and the lines attached to the dots in the figures represent the direction of motion. (Reproduced with permission from de Jong *et al.*, 1994. Copyright (1994) Oxford University Press.)

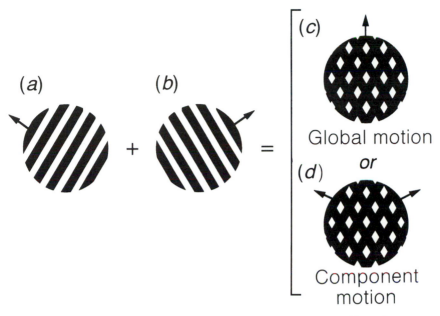

Fig. 10.4. Moving plaid patterns are achieved by the superimposition of two drifting gratings moving in different directions (*a* and *b*). Neurons in V1 are responsive to component motion, that is the direction of movement and speed of one of the two gratings (*d*). Neurons in V5 are responsive to global pattern motion, that is the direction of movement and speed of the whole pattern (*c*) (Redrawn from Stonor & Albright, 1993.)

screen in different directions, a human observer will perceive a single plaid pattern moving across the screen in a direction that is intermediate between the directions of the two gratings that make up the pattern (Fig. 10.4). We do not see the movement of two separate gratings (*component motion*), but rather see the *global motion* of the pattern. Single-unit recording has shown that neurons in layer 4B of V1 respond to component motion, but many neurons in V5 respond to global motion (Movshon *et al.*, 1985). That is to say, the responses of neurons in V5 correspond to our perception of motion. Fig. 10.5 displays another example of component and global motion. The edges of the square, viewed in isolation, seem to move in different directions, but the global motion of the square is in a single direction. In order to perceive the motion of objects in our environment, it is necessary to make the jump in processing from component to global motion sensitivity.

Using single-cell recording techniques, Thomas Albright (1984) mapped the characteristics of movement-sensitive neurons in area V5. Neurons with similar preferences seem to be grouped together in columns running per-

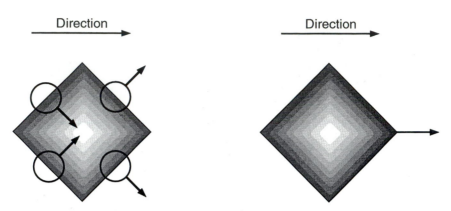

Fig. 10.5. A schematic diagram showing a moving square, detailing the difference between component motion on the right and global motion on the left. (Redrawn from Movshon *et al.*, 1985.)

pendicular to the surface of the cortex. Consistent with the organisation of many, if not all, visual areas, V5 is believed to be divided into modules or super-columns (Fig. 10.6). Each module consists of a pair of rectangles, arranged side by side. Moving along the long axis, one encounters neurons with directional sensitivities that vary systematically in a clockwise or counterclockwise fashion. The neurons in adjacent portions of each rectangle have motion sensitivities oriented in the opposite direction.

Bill Newsome and his colleagues trained a monkey to discriminate the direction of motion of a group of moving dots. The difficulty of the task could be varied by altering the proportion of dots moving coherently to the proportion of dots moving in a random direction (Fig. 10.7). Macaque monkeys were able to detect the direction of global motion when only 2–3% of the dots were moving in the same direction (Newsome & Paré, 1988). Newsome then carried out an ingenious experiment on V5 using chemical lesions. He destroyed V5 on the left side of the brain and left it intact on the other side. As a result, the intact V5 was able to act as a control for the lesioned V5. The crossover of the axons from the retinal ganglion cells at the optic chiasm means that the right side of the visual field is processed by the left hemisphere and the left visual field by the right hemisphere. In the right visual field of a monkey with a left hemisphere lesion, the detection threshold for global motion was raised by a factor of ten, relative to the unlesioned side (Newsome & Paré, 1988). The monkeys' ability to perceive stationary objects and stimuli was unaffected. Lesions to V5 do not abolish all motion sensitivity, as there are other parallel pathways processing motion information, such as through V3.

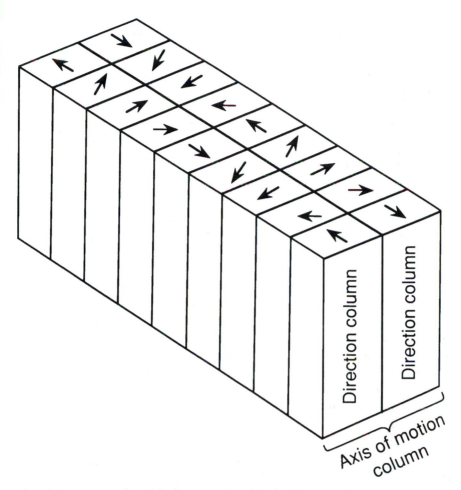

Fig. 10.6. Proposed model of a cortical module for direction-selective neurons in monkey V5. (Redrawn from Albright, Desimone & Gross, 1984.)

The behaviour of a single, directionally selective motion cell in V5 has been used to predict the behaviour of the whole animal on Newsome's motion discrimination task (Britten *et al.*, 1992). In each session, a V5 neuron was isolated with a microelectrode and the cluster of dots was positioned on its receptive field; the speed of the dots was matched to the cell's preferred speed. The dots either moved in the neuron's preferred direction or in the opposite direction, and the monkey was required to indicate in which direction the dots moved. The monkey's performance improved as the proportion of dots moving coherently increased. This improvement in performance was paralleled with an increase in strength of the neuron's response if the dots were

(a) 0%   (b) 50%   (c) 100%

Up

Down

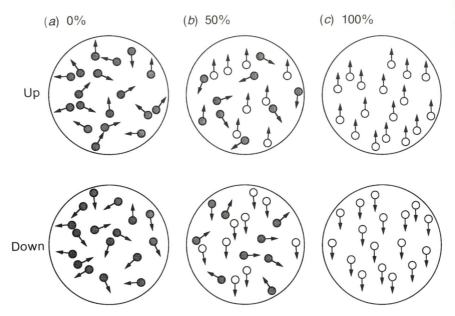

Fig. 10.7. Schematic representation of a random-dot stimulus that can be used to motion thresholds. In each panel, the arrow indicates the direction in which the attached dot moves. The dots moving in a random direction are shown in grey and those moving in the same direction are shown in white. (a) The dots are moving randomly and there is 0% correlation between their movements. (c) All the dots are moving together in the same direction and there is 100% correlation of movement. (b) Half the dots are moving in a random manner and half the dots are moving in the same direction. In this case, the dots are 50% correlated. (Redrawn with permission from Sekuler & Blake, 1994. Copyright (1994) McGraw-Hill.)

moving in its preferred direction. For roughly half the neurons tested, the proportion of dots showing coherence at which the neuron reliably signalled the correct direction closely matched the monkey's behavioural threshold. This is not to suggest that the monkey's decision is based on the response of a single cell, rather that the V5 neurons contribute to a population or ensemble of neurons. The activity of these populations form the basis from which the monkey's behaviour was computed.

Furthermore, microstimulation of small groups of neurons with the same or similar direction preferences alters the performance of the monkeys on the discrimination task (Salzman & Newsome, 1994). For example, if an electric current is passed through the microelectrode, which had previously been used to record from a cell, that cell is stimulated to respond as if its preferred visual stimulus had appeared in its receptive field. On the discrimination task, stimulating neurons with a preference for downward motion increased the probability of the monkey signalling that it had seen a downward movement, even when the dots were moving at random. The effect increased for currents of up to 40 mA but tended to reverse for larger currents (Murasagi, Salzman & Newsome, 1993). This has been interpreted as suggesting that larger currents activate neurons outside the targeted column, which have different direction preferences.

V5 seems to be divided into two subdivisions that analyse different aspects of motion, which seem to be related to two broad areas of function. These are the motion of an object through the environment and the motion effects caused by our own movement through the environment (Van Essen & Gallant, 1994). These subdivisions project to separate visual areas within the parietal lobe; medial superior temporal (MST) divisions l and d (MSTl, MSTd). Neurons in MSTl seem to be responsive to the motion of an object through the environment, whereas neurons in MSTd seems sensitive to motion caused by movement of our eyes or of ourselves. Neurons in this latter area are responsive to changes in certain parameters of a stimulus, such as an increase or decrease in its size (such as might be produced by moving toward or away from it), its rotation (such as might be produced when tilting our heads) and shear (such as might be produced when moving past objects at different distances) (Saito *et al.*, 1986; Duffy & Wurtz, 1991; Orban *et al.*, 1992). In addition, some cells are sensitive to mixtures of these stimulus parameters, such as spiral motion patterns, which have components of both rotation and expansion (Graziano, Andersen & Snowden, 1994). Therefore, cells in MSTd seem ideally suited to encode the visual changes that occur when we move and allow us to interact with our environment.

Human V5

Recent developments in non-invasive techniques for the study of brain function have allowed the position of V5 in humans to be mapped. Semir Zeki combined the techniques of PET and MRI to analyse the position of V5 in 12 normal subjects (Watson *et al.*, 1993b). The PET scanning was used to determine the position of areas of increased cerebral blood flow produced when subjects viewed a moving checkerboard pattern, compared with viewing the same pattern when it was stationary. The position of V5 based on PET was then compared with an image from the same brain obtained by MRI, allowing the position of V5 to be related to the gyral configuration of individual brains. The exact size and shape of the brain varies from individual to

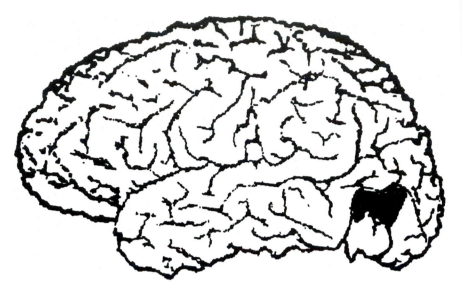

Fig. 10.8. The location of human V5 in the brain, based on PET and MRI scans. (Reproduced with permission from Zeki, Watson & Frackowiak, 1993 Copyright (1993) Oxford University Press.)

individual, and this is reflected in the position of V5, which can vary by 18–27 mm. However, there is a consistent relationship between the position of V5 and the sulcal pattern of the occipital lobe. V5 is situated ventrolaterally, just posterior to the meeting point of the ascending limb of the inferior temporal sulcus and the lateral occipital sulcus (see Fig. 10.8). This position has been confirmed by fMRI studies and by histological studies of post-mortem human brains (Tootell *et al.*, 1995b; Clarke, 1994).

Damage to human V5 seems to produce a similar set of deficits as lesions of V5 in monkeys. In 1983, a 43-year-old, female patient (L. M.) with brain damage was reported who showed a severe impairment in the perception of motion (*akinetopsia*), although otherwise her visual perception seemed normal (Zihl, von Cramon & Mai, 1983). She had normal acuity, stereo and colour vision and had no impairment of visual space perception or of the visual identification of shapes, objects or faces. A high resolution MRI scan showed that L. M. has bilateral damage to V5 (Shipp *et al.*, 1994). Although L. M. does not have complete loss of motion perception and can detect the presence of slowly moving objects, her perception of motion is generally impaired and she describes moving objects as undergoing episodic shifts in location. A graphic example is her perception of water being poured into a glass from a glass jug. She can see the glass and the jug but cannot see the water and the change in

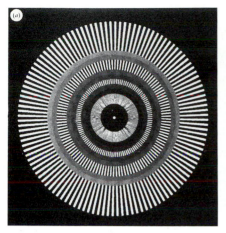 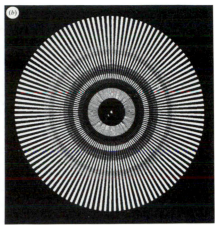

Fig. 10.9. The illusion is called Leviant's enigma. Fixation of the centre will result in the perception of rotatory motion in the circles. (Reproduced with permission from Zeki, 1994. Copyright (1994) Royal Society.)

water level in the glass until water has stopped being poured. A similar effect can be produced by temporarily inactivating V5 in normal human subjects using *transcranial magnetic stimulation* (*TMS*). In this process, a magnetic field is used to induce an electric current in a specific area of the brain. This temporarily inactivates the brain area, and when this technique is applied to human V5 the result seems to be complete akinetopsia (Beckers & Zeki, 1995).

Interestingly, activity in human V5, like that of monkey V5, can be correlated with perception of motion effects by the individual. The M pathway is largely colour blind. If a subject views a moving grating made up of alternating red and green stripes, as long as the stripes differ in luminance, motion will be perceived. However, if the two colours are equiluminant and the only cues to the existence of moving stripes are colour differences, then the perception of motion is strongly reduced. These changes correlate with activity in V5. Single-unit recording in monkeys and fMRI studies in humans have shown a reduction in the activity of V5 when the two colours become equiluminant (Tootell *et al.*, 1995a). The activity of V5 can also be correlated with the perception of motion in visual illusions. In illusions such as the enigma illusion (Fig. 10.9) or the waterfall illusion (see Chapter 5), there is no actual motion in the stimulus, but a human subject perceives movement as occurring. When the subject perceives motion, whether by actual movement or through illusion, V5 is active (Zeki, *et al.*, 1993; Tootell *et al.*, 1995a).

Transparent motion

Objects in the visual field are constantly passing in front of one another, yet the brain has no difficulty in distinguishing them. Neurons in V5 are sensitive to the speed and direction of moving objects, and each neuron has a preferred direction of motion that stimulates its maximal response. Many of these neurons are inhibited by motion in the opposite direction, and this inhibition is believed to help reduce noise and ensure an accurate representation of the moving stimulus. But many neurons also respond to stereoscopic disparity, which corresponds to the visual plane in which the stimulus occurs. A possible reason for the integration of these two apparently unconnected stimulus parameters has arisen from the work of Richard Andersen and his colleagues (Bradley, Qian & Andersen, 1995). They recorded from neurons in V5 of rhesus monkeys trained to fixate on a display screen. Individual neurons were first stimulated by patterns of dots moving in their preferred direction. Having confirmed the inhibitory effect of adding further dots moving in the opposite direction, Andersen and his colleagues separated the two sets of dots into two different planes (using coloured glasses similar to those for watching three-dimensional movies). They found that the inhibitory effect is strongest when both sets of dots lie in the same visual plane, and that it becomes weaker as the disparity increases. Similar results are obtained whether the two patterns overlap or are merely adjacent. The results suggest a simple explanation for how and why V5 integrates direction and depth. *Transparent motion* normally arises when objects pass each other in different planes. By not having inhibition between movement in different planes, the brain can interpret the two movements as independent. If, instead, the two movements occur within the same plane, transparent motion is less likely, and the brain looks for other explanations (such as random noise).

In fact, for simple patterns such as two sets of dots moving past each other, subjects have no difficulty in perceiving two sets of coherent motion, even when both occur in the same plane. But for more complex stimuli, the task becomes more difficult and stereoscopic disparity between the two directions can improve performance. In the real world, V5 may exploit not only depth but also other features, such as colour or texture, to distinguish between the components of transparent motion.

Key points

1. The structure and organisation of the retina means that only the central 2 degrees of the visual field are fully elaborated. To produce our perception

of the world, the eyes have to be constantly moving. There are two types of involuntary eye movement: micro saccades and saccades.

2. Micro saccades are small movements of the eye that destabilise the image on the retina and prevent the photoreceptors in the retina from adapting to a continuous stimulus.

3. Saccades are short, jerky movements of much larger size than micro saccades. These are used by the eye to explore the visual environment. During saccadic movement there seems to be a suppression of activity in the M pathway, which is normally sensitive to motion.

4. A third form of eye movement is called pursuit eye movement. This is under voluntary control and allows us to track moving objects.

5. Perception of external motion is analysed mainly within the M pathway. An important stage in this pathway is V5, whose neuronal responses to moving stimuli can be correlated with our own perception of these stimuli. The neurons in this area are arranged in a columnar and modular fashion, as seems to be the case in most, if not all, visual areas.

6. Damage to, or temporary inactivation of, V5 in humans causes deficits in the perception of motion (akinetopsia). Human V5 is active when we perceive motion, whether this perception is caused by real movement or by illusion.

11

Brain and space

The final frontier

The perception of depth is crucial to the generation of a three-dimensional representation of the spatial relationships in our surroundings, a representation that is essential if we are to be able to interact with our environment in any meaningful way. The visual system has two sets of depth cues: oculomotor and visual (Fig. 11.1). They are termed cues because they must be learnt through association with non-visual aspects of experience. Oculomotor cues are based on the degree of convergence (a measure of the angle of alignment) of the eyes and the degree of accommodation (change in shape) of the lens. The visual cues can be both monocular and binocular. The monocular cues include interposition, relative size, perspective and motion parallax. Binocular cues are based on the disparity between the different views of the world from the two eyes. From this disparity, a three-dimensional, or stereoscopic, representation can be generated. The information on depth, together with information about movement and velocity, seem to be integrated with information from other sensory modalities to produce a map of perceptual space that is common to all our senses. This integration seems to occur in the posterior parietal cortex. Damage to this area causes profound impairments in our perception of space, including that occupied by our own bodies.

Oculomotor cues

When you fixate an object, your eyes are accommodated and converged by an amount dependent on the distance between you and that object (Fig. 11.2). To

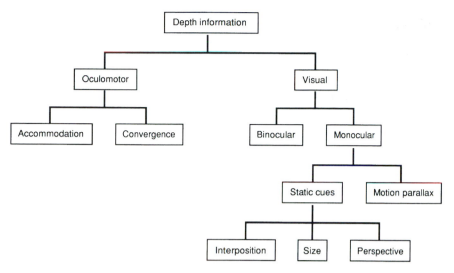

Fig. 11.1 Major sources of depth information. (Redrawn from Sekuler and Blake, 1994).

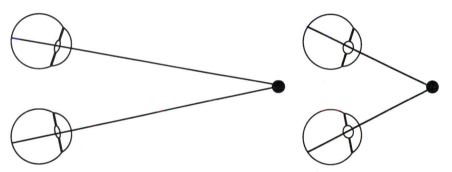

Fig. 11.2. An illustration of oculomotor cues. When objects move closer the lens thickens (accommodation) and the eyes turn together, changing their angle of alignment (convergence).

be seen clearly, close objects need more accommodation and convergence than do objects further away. By monitoring the degree of muscle strain, it is possible to determine values for the angle of convergence and the amount of accommodation. When you fixate an object more than a few metres from you, the muscle controlling accommodation is in its most relaxed state. So as a potential depth cue, accommodation would be useful only within the region immediately in front of you. Even within this region it is very inaccurate. This is true of most binocular vertebrates. The most striking exception is the chameleon (*Chameleo jackson*). The state of focus of the eye's lens acts as the principal distance cue; not only is the focusing mechanism of the chameleon's lens very fast and its range unusually extensive, it is also accurate enough to supply the muscles of the projectile tongue with all the information they need to shoot the sticky tip exactly the right distance to catch insect prey (Ott & Schaeffel, 1995). Part of the reason for this reliance on monocular cues may come from the way the chameleon's eyes seem to move independently and so limit the use of binocular cues.

The use of convergence can only operate over a limited range. The convergence angle formed by your two eyes vanishes to zero (because your eyes are looking straight ahead) when you are looking at objects more than 6 m away. Below this distance, convergence can be used as a reliable depth cue.

Interposition

When one figure occludes part of another (*interposition*), the partially occluded object is perceived as the more distant. The use of these cues develops such that by seven months of age human infants can judge relative distance solely on the basis of interposition. Young children often use interposition in their simple drawings, even though unable to reproduce any other pictorial depth cues. There is evidence that children as young as seven months determine distance solely on the basis of interposition (Yonas, 1984). Brain damage can remove the ability to use this cue, while leaving the ability to use other depth cues intact (Stevens, 1983).

A striking example of interposition is the Kanizsa triangle (see Fig. 8.2, p. 116), where distinct edges seem to be found where none exist. The visual system appears to employ a kind of interpolation process, where separate edges and contours in the same spatial neighbourhood are perceived to be connected if a connection can be formed by a simple line or curve and if the operation is consistent with the principle of interposition. This interposition is accomplished by neurons in V2 (Peterhans & von der Heydt, 1991). Although neurons in both V1 and V2 can respond to illusory contours defined by the co-linearity of line terminations, only neurons in V2 responded to illusionary contours extending across gaps. Additionally, even though we are unable to see the portions occluded, we assume the occluded objects are complete and do not have the occluded part missing. This process, which is called *amodal completion*, is also mediated by interpolation.

Relative size

As the distance between the viewer and an object varies, the size of the image of that object varies on the retina. This is true of any object viewed from any angle. So if the viewer is familiar with the size of an object, the size of its retinal image can be used to judge how far away the object is. Also, if a second object appears in the vicinity of the familiar object, the size of the second object can be judged by reference to the size of the familiar object. This scaling effect is used by building companies in show homes. The furniture is specially built to be 10% smaller than the standard size, and as the viewer scales the proportions of the unfamiliar rooms with reference to the relative size of the furniture, they will overestimate the size of the rooms. It helps give an impression of size and spaciousness seldom found in modern buildings and, more importantly as far as the builders are concerned, helps sell the houses.

Perspective

Perspective refers to changes in the appearance of surfaces of objects or surfaces as they recede into the distance. There are four forms of perspective cue. *Linear convergence* refers to the way parallel lines seem to converge with distance and is a way of giving the impression of depth in pictures and illustrations (Figs. 11.3 and 11.4). Another form is called *texture gradients*. Most surfaces have a texture, and the density of this texture will appear to increase with viewing distance. Texture gradients can, therefore, provide information on the distance and slope of surfaces, as well as information on the size of objects located on those surfaces. Moreover, rapid changes in texture gradients can signal the presence of edges or corners. The third form of perspective cue is called *aerial perspective*. This refers to the way an object further away seem less clear than those close up. This is the result of the scatter of light as it travels through the atmosphere, which has the result of reducing the contrast of the image (O'Shea, Blackburn & Ono, 1993). The degree of scatter is dependent on two factors: the distance between the object and the observer and the medium through which the light passes. For example, if the air contains a lot of dust or a lot of moisture droplets such as in a fog, more light will be scattered. The final perspective cue is *shading*. In the natural environment, light always comes from above, and so the pattern of shading can be used to derive depth (Ramachandran, 1988).

Fig. 11.3.   A size illusion, called the Ponzo illusion. Both bars are the same length, but the upper bar appears longer. This is because the perspective lines create the illusion that the upper bar is further away.

Motion parallax

As you move in the environment, the objects around you are constantly altering their position within your visual field. If you are travelling in a car or train, and you fixate a particular object within the passing scene, the pattern of relative movement around this object will demonstrate a phenomenon called motion parallax. Objects closer than the one you have fixated will appear to move in the opposite direction to yourself, whereas more distant objects will appear to move more slowly but in the same direction as yourself. The relative apparent motion of objects within your field of view as you move (*motion parallax*) provides a strong cue to the relative distance of objects from the observer.

Stereopsis

At close range, animals with overlapping visual fields have stereoscopic information available to them from the disparate images obtained at the two

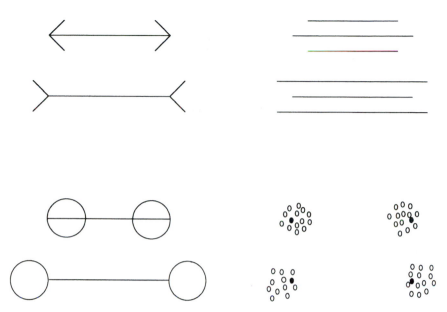

Fig. 11.4.   Four different forms of the Müller–Lyer illusion. The example on the top left is the traditional form. The two bars are of the same length, but the lower bar appears longer. In the example shown at the bottom right, the distance between the two black dots is the same in both lines, but the separation of the lower pair seems greater than that of the upper pair. Richard Gregory has suggested that the Müller–Lyer illusion could arise from implied depth information. The arrowheads on the ends of the vertical lines can be seen as angles formed by two intersecting surfaces. When the arrowheads point outwards, the two surfaces are seen to be slanted towards you. When the arrowheads point inwards the surfaces will be seen to be receding away from you. The perspective cues make the receding corner appears further away than the approaching corner. As the retinal image of the length of the two intersections is the same, it suggests that the receding line must be longer. However, the lower two forms of the illusion are difficult to explain in terms of the Gregory's theory, and it may be that his explanation needs to be reviewed.

eyes. Each of the eyes sees a slightly different view of the world as the result of the horizontal separation of the two eyes. The disparity is generated in the following manner. If the two eyes are focused on an object or point (B), the images in each eye are said to lie on corresponding points on the two retinae (Fig. 11.5). The images cast by a nearer or more distant object or point (C and A) will fall on disparate points on the two retinae; the amount of disparity will depend upon the distance between A and B and between B and C. If the brain can compute this disparity it will give precise information about the relative position of objects in the world. *Stereopsis* (whose literal meaning is solid appearance) requires binocular retinal disparity of some elements of a visual

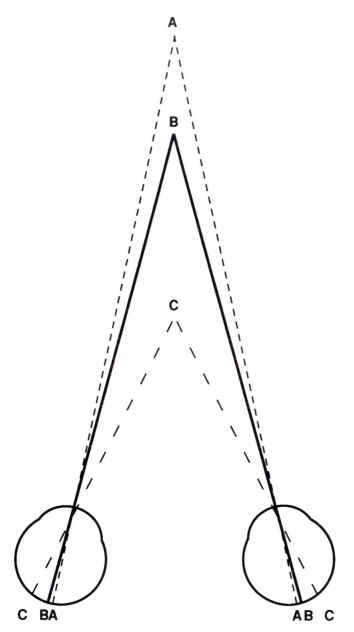

Fig. 11.5. The basis of visual disparity. Both eyes fixate at point B, so that the image falls on the fovea of both eyes. The images of point A (which is further away than B) and point C (which is closer) fall on non-corresponding points. These differences in where the two images are focused on the retina are called retinal disparities. The disparities can be used to calculate stereoscopic depth or stereopsis.

stimulus. The disparity effect can easily be demonstrated using a *stereoscope*, which was first developed by Sir Charles Wheatstone in 1838. This device presents two drawings (which together constitute a stereogram) separately to the two eyes. The two images are seen as one, and appear to be three-dimensional. Wheatstone's results demonstrate that reintroduction of retinal disparity gives the appearance of a three-dimensional image.

Disparity is calculated early in the visual pathways. Cells have been found in V1 of both monkeys and cats that respond maximally when their optimal stimuli fall on disparate areas of the two retinae (Clarke & Whitteridge, 1978; Poggio & Poggio, 1984). A cell selective for disparity responds most strongly to a stimulus lying at a particular distance. V1 contains two classes of neuron that seem to be sensitive to retinal disparity (Poggio & Poggio, 1984). These are binocular cells in layer 4B. The first of these classes of neuron responds with an increase or decrease in firing rate to a limited range of retinal disparities ($+0.2$ degree for the excitatory cells and $+0.4$ degree for the inhibitory cells). The second class of neuron selectively responds to stimuli nearer than, or farther away from, the fixation plane (the horopter). The biological significance of these cells receives support from the finding that there are two classes of person who have difficulties judging the distance of objects using binocular cues. Some of the first group misjudge objects in front of the fixation plane (closer to them), others have difficulties with objects that are behind it. The second group can easily judge which of two objects is farther from them when both objects are almost the same distance but have trouble judging which is closer when one object is much closer to them than the other.

The neural basis of three-dimensional space representation

Depth and motion cues can be used, in conjunction with other sensory cues, to produce a three-dimensional representation of our environment. The posterior parietal cortex (PPC) is the most likely site where such spatial relationships could be represented in the brain. PPC cells receive visual, auditory, somatosensory and vestibular sensory inputs, and the integration of these inputs could be used to form a map of perceptual space. This is a map of the location of objects around us with respect to ourselves that is common to all our senses.

The PPC is composed of the *superior* and *inferior parietal lobules*. In humans, damage to the superior parietal lobule causes deficits in complex somaesthetic judgements. These deficits can manifest themselves in several ways. A subject may be unable to recognise the shape of objects by touch alone (*astereognosis*).

This symptom is often linked to a more general deficit of body image, where a subject is unable to assimilate spatial impressions of the positions of his/her limbs and body to build up an accurate body image (*amorphosynthesis*). A related disorder is *asomatognosia*, which is the denial of the existence of a part of the body. A patient suffering from this condition may deny that one of his limbs belongs to him or her.

Damage to the inferior parietal lobule is associated with disruption of spatial vision and spatial orientation. Symptoms include deficits in reaching and pointing to visual targets, avoiding obstacles, learning and remembering routes, judging distance and size, recognising spatial relations, fixating a target and following a moving stimulus. However, such subjects usually have normal object vision. A particularly unusual effect, called visual neglect, is found with damage to the inferior parietal lobule on just one side of the brain.

Damage to early visual areas, such as V1, cause scotomas (holes in our visual perception). Damage to higher association areas in the 'what' pathway, such as IT in monkeys or the fusiform gyrus in humans, impairs the ability to recognise objects (agnosias). Damage to the inferior parietal lobule does not produce scotomas in our visual field but has a more subtle effect. Small lesions one side of the parietal cortex reduce the accuracy of localising objects on the contralateral (opposite) side. The effect is most marked with lesions of the right side. As the size of the lesion increases, so does the magnitude of effect, until at a certain point the whole of the contralateral side is ignored or *neglected*. For example, patients with unilateral lesions of the inferior parietal lobule often dress or shave only one side of the body, draw one side of a picture and attend to only one half of space, both near and far (Andersen, 1989). Unilateral neglect can also extend to imagery and memory. For example, Milanese patients with right parietal lesions were asked to imagine they were standing at one end of the Piazza del Duomo facing the cathedral and describe from memory the buildings along the sides of the piazza. The patients described only the buildings on the side contralateral to the intact parietal lobe (Bisiach & Luzzatti, 1978). However, when asked to imagine they were standing on the other side of the piazza with their backs to the cathedral, they then described the opposite set of buildings. Similarly, parietal patients suffering from neglect of the left side of space were unable to spell the beginnings of short words, as if spelling involved reading from an imaginary screen, the left side of which was faded (Baxter & Warrington, 1983).

Bisiach, Capitani & Porta (1985) tested whether the boundary of neglect moved with the retinal field or whether it was anchored to head or body. Their conclusion was that their patients used at least two co-ordinate systems, one relating to the body axis and the other relating to line of sight (oculomotor).

Hence, neglect displayed by patients with damage to the inferior parietal lobule is seldom purely retinotopic. The space they ignore does not move each time they move their eyes but tends to be centred on a point passing through the centre of the body or head (the egocentre).

In humans, left neglect is much more common than right. Hemispheric specialisation has lead to a concentration of visuospatial functions in the right PPC. Since lesions of the left PPC seldom give rise to neglect, it seems likely that the right PPC duplicates partially the spatial functions of the left PPC for the right hemifield. This inference is supported by the fact that patients with right PPC damage and left neglect often show some degree of inattention to targets up to 10 degrees in their right (ipsilateral) field. This result is supported by the finding that in monkeys many parietal neurons have receptive fields that extend well into the ipsilateral field, suggesting that more neurons in the right PPC have bilateral receptive fields than neurons in the left PPC. John Stein has suggested that the right PPC has become specialised for representing three-dimensional space and the direction of attention within this space, and the left PPC has become specialised to directing attention to temporal order (Stein, 1992). Human attributes such as speaking, logic and calculation require the skill of being able to sequence events in time accurately, and these attributes are most impaired by lesions of the left side.

There is a suggestion that patients with unilateral neglect have a degree of 'unconscious' perception of objects within their neglected field, although they may deny this. For example, a patient with right parietal damage (and, there-fore, left field neglect) performed a forced-choice task in which he was asked to decide whether there were differences between two houses, one with flames coming out of the left-hand window. The patient's ability to discriminate between the two images was not above chance in this task, but if asked which house he preferred, he consistently chose the house without flames, suggest-ing an unconscious processing of the differences between the houses (Marshall & Halligan, 1988; Bisiach & Rusconi, 1990).

The representation of space in the PPC seems to be divided into three cat-egories: personal, peripersonal and extrapersonal space (Stein, 1992). *Personal space* is the space occupied by our own body. Its co-ordinates are based on the orientation of the head, signalled by the vestibular system, together with information about the position of the neck and limbs. This information is pri-marily represented in the superior parietal lobule. *Peripersonal*, or *near space,* is the space surrounding us within which we can reach out and touch objects. Its neural representation requires the integration of retinal foveal signals with oculomotor and limb movement information, which occurs within a subdivi-sion of the inferior parietal lobule. *Extrapersonal*, or *far space*, is the space beyond peripersonal space. Its representation requires the integration of visual, auditory and oculomotor cues and whole body signals. This again seems to be represented in a subdivision of the inferior parietal lobule. The three forms of spatial representation seem to be functionally and anatomically separate. For example, a patient has been reported with posterior parietal damage who exhibited left unilateral neglect within peripersonal space, but not in far space (Halligan & Marshall, 1991).

Key points

1. The perception of visual depth is based on both oculomotor and visual cues. Oculomotor cues are derived from changes in accommodation and convergence when the eye focuses on an object.
2. There are both monocular and binocular visual cues. The monocular cues include:
   (i) interposition, which is when one object partially occludes another and the occluded object is perceived as being more distant; the occluded objects are assumed to be complete, a process called amodal completion
   (ii) relative size, which is where smaller objects are assumed to be more distant than larger objects
   (iii) perspective, which is the change in the appearance of surfaces or objects as they recede into the distance
   (iv) motion parallax, which uses the pattern of relative movement around an object to derive a measure of depth.
3. Binocular cues are based on the slightly different views of the world that the two eyes see because of their horizontal separation. The difference in the two views is called disparity and it is used to calculate a three-dimensional or stereoscopic representation of an object.
4. It seems that depth and space is represented in the posterior parietal cortex, an area which is also important for spatial attention.
5. Damage to the superior parietal lobule produces an impairment of body image.
6. Damage to the inferior parietal lobule, especially on the right, produces a phenomenon in which the whole of the contralateral side is ignored or neglected.
7. The representation of space is divided into three categories: personal, peripersonal and extrapersonal space. Personal space is the space occupied by your body, peripersonal space is the space within reach and extrapersonal space is the space beyond peripersonal space.

12

Building the visual image

Putting it all together

As we have seen in previous chapters, visual information is broken down into its components and processed, in parallel, in specialised areas. Cells in different areas show a preference for different combinations of, for example, colour, motion, orientation, texture, shape and depth. This is all carried out in a complex network of 32 visual areas connected by at least 305 connections (Van Essen *et al.*, 1992). These connections can run in three 'directions'. First, from lower areas (such as V1) to higher areas (such as V2). These are called feedforward connections. Second, all these feedforward connections have reciprocal feedback connections running from higher to lower areas. Third, there are also lateral connections running from areas of equivalent processing complexity. It seems a far from trivial task to reintegrate all of this information from this complex network into the seamless, coherent perception of the world we all share. There are two obvious problems. First, we have to put all the different stimulus features of an object back together in the right spatial and temporal relationship to one another. Second, we seldom see a single object in isolation. How, when we are dealing with the perception of two or more objects simultaneously, do we differentiate which features belong to which object? Failure to carry out this task correctly is called the superposition catastrophe.

In 1981, Rudolf von der Malsburg noticed a similar problem to those outlined above with neural network models. These simulations had the serious drawback that they could only represent the presence of two or more features in their input, having no means by which to segregate and bind the different features of two objects that were presented at the same time. The different features of the two objects would, therefore, be confused, producing non-existent feature conjunctions. Von der Malsburg suggested that a binding mechanism was required, by which the different features of different objects could be separately associated. He suggested that the timing of the neural action potentials or spikes could be used in a temporal code to show which features were associated with which object. The idea of a temporal code seemed like a good idea: temporal relations between the firing of sensory neurons might provide a solution to the problem of binding the activity distributed both between and within cortical areas into coherent representations. The case for a temporal code in feature binding received considerable support when experimental results on neuronal oscillations in cat visual cortex were reported.

Neuronal oscillations

Two research groups in Germany made microelectrode recordings from the visual cortex of cats and found that neurons in several visual areas of the cat exhibited short-duration periods when their activity was oscillatory (Eckhorn *et al.*, 1988; Gray & Singer, 1989). These oscillations were said to be 'stimulus-related' because the frequency spectrum during stimulation showed peaks between 30 and 70 Hz, in contrast to the low-frequency peaks (1 to 30 Hz) of background activity. It was suggested that these phenomena related to von der Malsburg's idea of a temporal code: neurons responding to different features of the same object could oscillate in phase, while cells responding to the features of different objects would oscillate out of phase. Both groups demonstrated that synchronisation of oscillatory activity could occur (Eckhorn *et al.*, 1988; Gray *et al.,* 1989; Gray & Singer 1989) and showed also that the likelihood of phase coherence declined with the separation of the two electrodes and with the dissimilarity in the orientation preferences of the recorded neurons (Gray *et al.,* 1989). These characteristics of phase coherence, however, could simply reflect structural properties of the underlying cortical network of connections, and it remains to be demonstrated that phase relations between oscillating cells could, as predicted, reflect the degree to which the cells are responding to the same object.

Tests of temporal binding in the cat

The existence of oscillatory activity is not proof that these oscillations function as part of a temporal binding mechanism. To provide evidence for this hypothesis, different research groups have made microelectrode recordings in the cat visual system to determine whether features of the oscillatory activity (such as their frequency and whether different populations of neurons synchronise) alter in a manner consistent with a role in temporal binding.

There are as many feedback connections between areas as there are feedforward connections. In cats, the feedback projections may play a role in synchronising the responses of cells projecting to the primary visual cortex. Some LGN cells are reported to show a greater degree of response synchronisation when stimulated by visual stimuli than would be expected if the responses of the cells were independent (Sillito et al., 1994). The synchronisation is between cells activated by the same visual stimulus feature, in this case a moving light bar. If the same pair of cells is co-activated by two spatially segregated, simultaneous light spots, their responses do not synchronise. Moreover, this synchronisation is abolished by the removal of the visual cortex, which suggests that it is generated by the thalamic feedback projection. Simple cells in V1 receive converging inputs from LGN cells to produce their responses to lines and edges (see Fig. 5.1, p. 80). Singer has suggested that synchronisation is established when an array of LGN cells is activated by a contour whose orientation matches the preference of those cortical cells that receive convergent input from that array of LGN cells (Singer, 1994). In short, it is suggested that feedback from V1 labels those LGN cells signalling a particular feature, such as a line or edge.

The temporal binding theory predicts that the phase relations between oscillating cells reflect the degree to which the cells are responding to the same object. The lack of overlap of the receptive fields of spatially separated cells has been exploited to test this important prediction. On two occasions, phase coherence between two recording sites with similar orientation preference was reported to be best when their receptive fields were stimulated by a single long bar (Gray et al., 1989). But it was less strong when the two receptive fields were stimulated by two short bars moved in the same direction, as would be consistent with a single bar moving behind an object. Oscillatory synchronisation was absent when the two short bars were moved in opposite directions (Gray et al., 1989). The reduction in phase coherence during presentation of a partially occluded bar compared with that during presentation of an unoccluded bar is surprising; this is just the situation when, according to the theory, a binding mechanism would be expected to operate, yet the proposed binding mechanism was weakened in this situation.

Another important prediction of the oscillation-binding theory was tested in a further report that concerned cells with overlapping receptive fields but different orientation preferences (Engel et al., 1991). A single light bar, whose

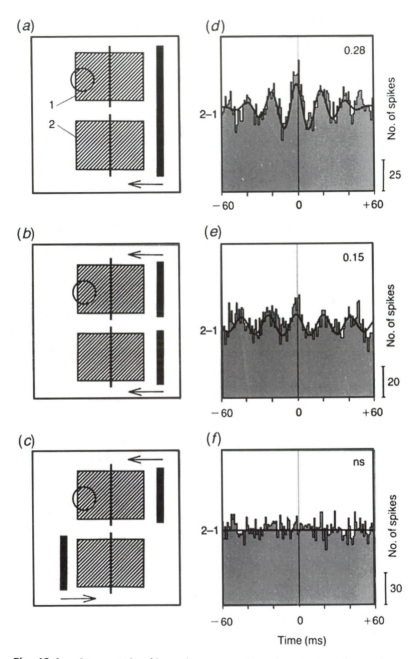

Fig. 12.1. An example of how the proposed synchronisation of neural responses are influenced by visual stimuli. Multi-unit activity was recorded from two sites in cat V1 that were separated by 7 mm. The cells responded most strongly when the stimulus bar was at a vertical orientation. (a, b and c) Plots of the receptive fields

orientation was mid-way between the optimal orientations for two sets of neurons, stimulated both sets of neurons and gave rise to phase-coherent responses when swept across the cells' receptive fields (Figs. 12.1 and 12.2). When two bars at the preferred orientations of the two neurons were moved across the receptive fields, both sets of cells were again stimulated but coherence was no longer apparent between cells with different orientation preferences. A more elaborate version of this experiment used four sets of neurons. The degree of coherence in the responses of neurons being dependent upon the orientation of the stimulus bar. This experiment was interpreted as evidence for the cells moving into different ensembles by changing the temporal relationships of their firing patterns. Like the 'long bar' experiment, above, this conclusion appears drawn from very few empirical examples: no other instance of four cells changing their relative temporal synchronisation in this way has been reported in detail, and no further reports of the phenomena in the 'long bar' experiment have appeared.

The methods Singer's group used to analyse their data have been criticised, and it has been suggested that they may have overestimated the proportion of neurons displaying synchronised oscillatory activity (for a detailed critique see Young, Tanaka & Yamane, 1992). In the last two experiments described above, the number of positive results is very low, and if the analysis technique is flawed, they may be false positives. Limited reanalysis of a subset of the data from the Singer's group, taking into account only some of these criticisms,

Caption for Fig. 12.1 (*cont.*)
(shown as hatched rectangles) under three different stimulus paradigms. (*a*) A long continuous light bar moving across both fields; (*b*) two independent light bars moving in the same direction; (*c*) the same two bars moving in opposite directions. The circle represents the centre of the cat's visual field, and the black line drawn across each receptive field indicates the preferred orientation of those cells. (*d, e* and *f*) Correlograms obtained with each stimulus paradigm. Using the long light bar, the two oscillatory responses were synchronised as indicated by the strong modulation of the crosscorrelogram with alternating peaks and troughs (*d*). If the continuity of the stimulus was interrupted, the synchronisation became weaker (*e*), and it totally disappeared if the motion of the stimulus was not coherent (*f*). The change in the stimulus configuration affected neither the strength nor the oscillatory nature of the two responses. The graph superimposed on each of the correlograms represents a Gabor function that was fitted to the data to assess the strength of the modulation. The number in the upper right corner indicates what Singer and his colleagues call the 'relative modulation amplitude,' a measure of the correlation strength that was determined by calculating the ratio of the amplitude of the Gabor function to its offset; ns, not significant. Scale bars indicate the number of spikes. (Reproduced with permission from Engel *et al.*, 1992. Copyright (1992) Elsevier Trends Journal.)

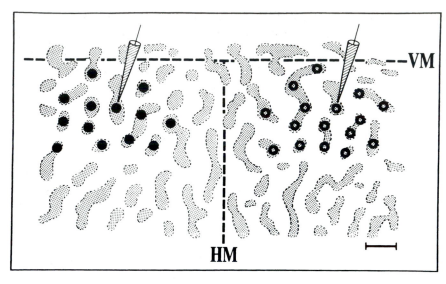

Fig. 12.2. A schematic of the long-bar experiment illustrated in Fig. 12.1. The location of the recording electrodes is mapped onto a representation of the orientation columns in cat V1. Shaded areas correspond to bands of orientation columns responding to vertical contours. Electrode symbols indicate the two groups of cells from which recordings were taken. The diagram illustrates the measurements in Fig. 12.1 c, f, where recordings were taken in response to light bars moving in opposite directions. The two cell groups were assumed to be part of two large assemblies (represented by filled circles and open circles), which are activated by two light bars. It is suggested that the members of each of the assemblies oscillate in synchrony, but there is no constant temporal relationship existing between them. The two assemblies are spatially separate. When the long continuous light bar is used, it is proposed that the intervening shaded orientation bands would also be activated, creating the formation of a single large assembly of synchronously oscillating groups. HM, representation of the horizontal meridian; VM, representation of the vertical meridian. The scale bar is 1 mm. (Reproduced with permission from Engel *et al.*, 1992. Copyright (1992) Elsevier Trends Journals.)

demonstrates significant overestimation of the number of sites showing phase coherence (Gray *et al.*, 1992). Therefore, the proof that synchronisation of oscillations can act as a temporal binding mechanism is far from conclusive. Moreover, while there is little disagreement that oscillating responses can sometimes be detected in cat visual cortex, there is considerable disagreement about the origins of oscillatory activity. Ghose & Freeman (1992) found that the strongest oscillatory signals were in the LGN and not in the cortex. This is an important distinction, as Singer considers the oscillation phenomena to be the result of a dynamic cortical generation process. Moreover, it is known that oscillations of the same frequency range can be found in the neural activity of cat retinal ganglion cells and in the optic nerve. Ghose and Freeman (1992) suggest that the oscillations measured in the visual cortex are just the product of this spontaneous activity in the retina passing through the visual system, with no functional significance at all.

Neural oscillations in the primate visual system

There is yet more disagreement when one turns to the issue of whether this oscillatory activity occurs in the monkey. In the macaque V1, one study showed that, with very similar conditions of anaesthesia and stimulation, the oscillations of the type reported in the cat were absent (Young *et al.*, 1992), but oscillations have been reported in the macaque V1 by Eckhorn and co-workers (1993) and by Livingstone (1991) in the squirrel monkey V1. The differences seen in the results of studies examining the existance of oscillations continue into the extrastriate cortex. Singer and his colleagues have reported oscillations in macaque V5 (Kreiter & Singer, 1992), but other groups found no such oscillations (Young *et al.*, 1992; Bair *et al.*, 1994). Nakamura, Mikami & Kubota (1992) reported oscillations in macaque IT, but two other groups failed to find any evidence of oscillations in this area (Young *et al.*, 1992; Tovée & Rolls, 1992).

Theoretical issues

Three aspects of the reported characteristics of the oscillations seem to be inconsistent with a binding mechanism that could function within the known parameters of the visual system. First, most studies have used dynamic stimuli, usually moving bars. In the cat, the reported oscillation frequency and amplitude are largely proportional to stimulus velocity (Gray *et al.*, 1990), and it seems that there are few or no oscillations in the response of visual cells to static stimuli both in cats and primates (Gray *et al.*, 1990; Tovée & Rolls, 1992). At best, this suggests a separate temporal binding mechanism for processing dynamic and static stimuli, which seems unnecessarily complicated and elaborate. At worst, it suggests that oscillations are merely an artefact of the processing of dynamic stimuli and have no functional relevance.

Second, for both cats and monkeys, the dynamic stimuli used to elicit neuronal oscillations had lower velocities or contrasts than those which evoke the strongest spike discharges from the cells they were used to stimulate (Ghose &

Freeman, 1992; Eckhorn *et al.*, 1993). Ghose and Freeman note that although oscillations were not detectable in the spontaneous activity of cortical cells, possibly because of their low firing rate, oscillations are detectable once the stimulus strength is raised just enough to stimulate firing. As the stimulus strength is increased (in this case the stimulus contrast) and the cell's firing rate increases, the oscillations become weaker. After all, in the case of the suboptimal stimuli, what are the oscillations binding? Moreover, is it likely that a population of coherently oscillating cells is signalling much less specifically than a population of maximally firing cells? Both contingencies seem remote, a less elaborate explanation of the results would be that the oscillations are an artefact related to certain stimulus parameters and are not involved in temporal binding.

The third and last inconsistency is based on the temporal properties of the oscillation spindles. For visual stimuli, the processing time at each synapse seems to be strictly limited, perhaps to as little as 10–20 ms (Tovée, 1994b). The application of Bayesian information theory to the analysis of spike trains from visual neurons in the primate temporal visual cortex suggested that up to 67% of the information available in a 400 ms sample of the spike train may be available in a 20 ms sample and up to 87% of the information may be available in a 50 ms sample (Fig. 12.3) (Tovée *et al.*, 1993; Tovée & Rolls, 1995). This result is consistent with the findings of a series of experiments comparing the time required for a human subject to recognise a stimulus with the functioning of visual neurons in the macaque temporal lobe (Rolls & Tovée, 1994; Rolls *et al.*, 1994). In these experiments, a brief target image was presented for only 20 ms, followed at a variable time interval (the stimulus onset asynchrony, SOA) by a mask stimulus. The mask stimulus had the effect of limiting the amount of time the visual cells in the temporal cortex had to respond to the target stimulus. At SOA values at which a human subject could just recognise a visual stimulus, visual cells in the temporal visual cortex that were optimally sensitive to this stimulus were active for only 20–30 ms. The evidence suggests that information in the visual system is processed and transferred to the next visual area very rapidly. Oscillations have been reported to occur in spindles that are not locked to the stimulus onset, develop over tens of milliseconds and persist for 100–300 ms in the visual cortex (Singer, 1993). It is hard to see how a binding mechanism with such temporal properties could play any part in the rapid visual processing which is sufficient for the recognition of static objects.

Is a temporal binding mechanism necessary?

The need for a temporal binding mechanism seems to be based on the assumption that the visual scene we perceive is based on a uniform input of

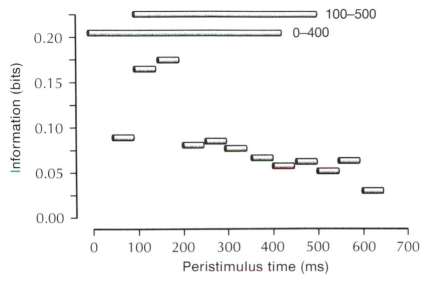

Fig. 12.3. The information available about which image was shown to a monkey, based on 50 ms samples of the spike train of face-selective neurons. The time period over which each sample was taken is indicated by the length and position of each horizontal line. For comparison, the information available in 400 ms samples starting at 0 ms and 100 ms post-stimulus onset are also shown. (Reproduced with permission from Tovée 1994b. Copyright (1994) Current Biology.)

information across the whole 180 degrees of visual field. However, in primates, high acuity is supported by only the central 2 degrees of the retina, based on the fovea. The packing of the cone cells, which mediate both high acuity and colour vision, declines rapidly with increasing eccentricity. The cone density decreases by a factor of 30 between the centre of the fovea and an eccentricity of 10 degrees (Curcio et al., 1991). The number of photoreceptors synapsing onto a single ganglion cell also increases with increasing eccentricity (Perry & Cowey, 1985). As a result the 'grain size' of the picture increases with increasing retinal eccentricity. This is also true of colour, where colour discrimination thresholds rise as an exponential function of eccentricity (Nagy & Wolf, 1993). These responses are reflected in the relative proportions of neural space given over to processing the input from the central and peripheral retina. The representation of the central visual field in the V1 is up to six times larger than that for the periphery (Azzopardi & Cowey, 1993). The cumulative result of this organisation is to create almost a tunnel vision effect, where only the visual information from the centre of the visual field is fully sampled and analysed. This is accentuated by the effects of attention. If a monkey's attention is focused on a stimulus, then the response of a visual cell to that stimulus is not enhanced, but the cell's response to other stimuli within it's receptive field is reduced (Moran and Desimone, 1985). The net effect of attention seems to be to reduce the functional size of a cell's receptive field.

So how are the different features of the visual input, limited though they are by the physical constraints of the visual system, coherently integrated within the cortex? Anatomical and functional modelling studies have shown that there is considerable reconvergence of processed visual information in the frontal lobe, the rostral STS and the limbic system (Van Essen *et al.*, 1992; Young, 1992; Young *et al.*, 1995). We have already seen how the response properties of visual neurons have become more and more complex. Therefore, integration could be brought about by hierarchical convergence, rather than by the synchronisation of populations of neurons spread widely in the brain.

Key points

1. Different features of a visual stimulus are analysed in different specialised visual areas. This raises the question of how all this distributed information is reintegrated, and how to prevent features from different objects being mixed together (the superposition catastrophe).
2. It has been suggested that a binding mechanism is required by which the different features of different objects could be separately associated. The timing of the neural action potentials or spikes could be used in a temporal code to show which features are associated with which object.
3. Neurons in cat visual areas exhibit short-duration periods when their activity is oscillatory (30–70 Hz). It has been suggested that these periods are part of a temporal code: neurons responding to different features of the same object could oscillate in phase, while cells responding to the features of different objects would oscillate out of phase.
4. Although there is general agreement that oscillatory activity does occur under some circumstances, the evidence for its synchronisation under conditions where it could function as a temporal binding mechanism is inconclusive.
5. There are certain theoretical inconsistencies in the use of oscillations as a temporal binding mechanism:
    (i) the frequency and amplitude of the oscillations seem to be proportional to the speed of the stimulus, and static stimuli do not seem to produce oscillations

(ii) the fact that suboptimal stimuli are required to produce measurable oscillations casts suspicion on the role of oscillations in the normal processing of visual information

(iii) visual information is processed very rapidly in the visual system, and the temporal features of the oscillations may not allow them to function as a binding mechanism under these conditions.

6. An alternative scheme for the integration of different processing streams is by hierarchical convergence on cells in a subsequent higher integrative cortical area, a hypothesis supported by anatomical, lesion and single-unit recording studies.

References

Adolphs, R., Tranel, D., Damasio, H. & Damasio, A. (1994). Impaired recognition of emotion in facial expressions following bilateral damage to the human amygdala. *Nature,* 372, 669–672.

Albright, T.D. (1984). Direction and orientation selectivity of neurons in visual area MT of the macaque. *Journal of Neurophysiology*, 52, 1106–1130.

Albright, T.D., Desimone, R. & Gross, C.G. (1984). Columnar organization of directionally selective cells in visual area MT of the macaque. *Journal of Neurophysiology*, 51, 16–31.

Allison, T., Begleiter, A., McCarthy, G., Roessler, E., Nobre, A.C. & Spencer, D.D. (1993). Electrophysiological studies of processing in human visual cortex. *Electroencephalography and Clinical Neurophysiology,* 88, 343–355.

Allison, T., Ginter, H., McCarthy, G., Nobre, A.C., Puce, A., Luby, M. & Spencer, D.D. (1994). Face recognition in human extrastriate cortex. *Journal of Neurophysiology*, 71, 821–825.

Amaral, D.G., Price, J.L., Pitkanen, A. & Carmichael, S.T. (1992). Anatomical organization of the primate amygdaloid complex. In *The Amygdala: Neurobiological Aspects of Emotion, Memory and Mental Dysfunction.* ed. J.P. Aggleton, pp. 1–66. New York: Wiley-Liss.

Andersen, R.A. (1989). Visual eye-movement functions of the posterior parietal cortex. *Annual Review of Neuroscience*, 12, 377–403.

Artola, A. & Singer, W. (1987). Long-term potentiation and NMDA receptors in rat visual cortex. *Nature*, 330, 649–652.

Asjeno, A.B., Rim, J. & Oprian, D.O. (1994). Molecular determinants of human red/green color discrimination. *Neuron*, 12, 1131–1138.

Awaya, S., Miyake, Y., Imayuni, Y., Kanda, T. & Komuro, K. (1973). Amblyopia in man, suggestive of stimulus deprivation amblyopia. *Japanese Journal of Ophthalmology,* 17, 69–82.

Azzopardi, P. & Cowey, A. (1993). Preferential representation of the fovea in the primary visual cortex. *Nature,* 361, 719–721.

Bair, W., Koch, C., Newsome, W. & Britten, K. (1994). Power spectrum analysis of bursting cells in area MT in the behaving monkey. *Journal of Neuroscience,* 14, 2870–2892.

Banks, M.S., Aslin, R.N. & Letson, R.D. (1975). Sensitive period for the development of human binocular vision. *Science*, 190, 675–677.

Barbur, J.L., Watson, J.D.G., Frackowiak, R.S.J. & Zeki, S. (1993). Concious visual perception without V1. *Brain*, 116, 1293–302.

Barlow, R.B. (1972). Single units and sensation: a neuron doctrine for perceptual psychology. *Perception*, 1, 371–394.

Barlow, R.B., Birge, R.R., Kaplan, E. & Tallent, J.R. (1993). On the molecular origin of photoreceptor noise. *Nature*, 366, 64–66.

Baxter, D.M. & Warrington, E.K. (1983). Neglect dysgraphia. *Journal of Neurology, Neurosurgery and Psychiatry,* 46, 1073–1078.

Baylis, G.C. & Rolls, E.T. (1987). Responses of neurons in short-term and serial recognition memory tasks. *Experimental Brain Research*, 65, 614–622.

Baylor, D.A. (1987). Photoreceptor signals and vision. Proctor Lecture. *Investigative Ophthalmology and Visual Science*, 28, 34–49.

Bear, M.F., Kleinschmidt, A., Gu, Q. & Singer, W. (1990). Disruption of experience dependent synaptic modifications in striate cortex by an infusion of an NMDA receptor antagonist. *Journal of Neuroscience,* 10, 909–925.

Bear, M. & Singer, W. (1986). Modulation of visual cortex plasticity by acetylcholine and noradrenaline. *Nature,* 320, 172–176.

Beckers, G. & Zeki, S. (1995). The consequences of inactivating areas V1 and V5 on visual motion perception. *Brain,* 118, 49–60.

Biederman, I. (1987) Recognition by components: a theory of human image understanding. *Psychological Review*, 94, 115–145.

Bindman, L. & Lippold, O. (1981). *The Neurophysiology of the Cerebral Cortex.* London: Edward Arnold.

Bisiach, E., Capitani, E. & Porta, E. (1985). Two basic properties of space representation in the brain – evidence from unilateral neglect. *Journal of Neurology, Neurosurgery and Psychiatry,* 48, 141–144.

Bisiach, E. & Luzzati, C. (1978). Unilateral neglect of representional space. *Cortex,* 14, 129–133.

Bisiach, E. & Rusconi, M.L. (1990). Breakdown of perceptual awareness in unilateral neglect. *Cortex*, 26, 643–649.

Blakemore, C. & Cooper, G.F. (1970). Development of the brain depends on the visual environment. *Nature,* 228, 477–478.

Blakemore, C. & Mitchell, E.D. (1973). Environmental modification of the visual cortex and the neural basis of learning and memory. *Nature,* 241, 467–468.

Blasdel, G.G. & Salma, G. (1986). Voltage sensitive dyes reveal a modular organization in the monkey striate cortex. *Nature,* 321, 579–585.

Bliss, T.V.P. & Collingridge, G.L. (1993). A synaptic model of memory: long-term potentiation in the hippocampus. *Nature*, 361, 31–39.

Bradley, D.C., Qian, N. & Andersen, R.A. (1995). Integration of motion and stereopsis in middle temporal cortical area of macaques. *Nature*, 373, 609–611.

Britten, K.H., Shadlen, M.N., Newsome, W.T. & Movshon, J.A. (1992). The analysis of visual motion: a comparison of neuronal and psychophysical performance. *Journal of Neuroscience*, 12, 4745–4765.

Brothers, L. & Ring, B. (1993). Mesial temporal neurons in the macaque monkey with responses selective for aspects of social stimuli. *Behavioural Brain Research*, 57, 53–61.

Bruce, V. & Green P.-R. (1990). *Visual Perception, Physiology, Psychology and Ecology.* Hove, UK: Lawrence Erlbaum.

Buisseret, D. & Imbert, M. (1976). Visual cortical cells: their developmental properties in normal and dark reared kittens. *Journal of Physiology*, 255, 511–525.

Burr, D.C., Morrone, M.C. & Ross, J. (1994). Selective suppresion of the magnocellular visual pathway during saccadic eye movements. *Nature*, 371, 511–513.

Carey, S. & Diamond, R. (1977). From piecemeal to configurational representation of faces. *Science*, 195, 312–314.

Cartier, M., Tsui, L.-P., Ball, S.P. & Lubsen, N.H. (1994). Crystallin genes and cataract. In *Molecular Genetics of Inherited Eye Disorders,* ed. A.F. Wright & B. Jay. Switzerland: Harwood Academic.

Castiello, U., Scarpa, M. & Bennett, K. (1995). A brain damaged patient with an unusual perceptuomotor deficit. *Nature*, 374, 805–808.

Clarke, P.G.H. & Whitteridge, D. (1978). A comparison of stereoscopic mechanisms in cortical visual areas V1 and V2 of the cat. *Journal of Physiology*, 275, 617–618.

Clarke, S. (1994). Modular organization of human extrastriate visual cortex: evidence from cytochrome oxidase pattern in normal and macular degeneration cases. *European Journal of Neuroscience*, 6, 725–726.

Corbetta, M. Miezin, F.M. Dobmeyer, S. Shulman, G.L. & Petersen, S.E. (1991). Selective and divided attention during visual discrimination of shape, color and speed: functional anatomy by positron emission tomography. *Journal of Neuroscience*, 11, 2383–2402.

Cornsweet, T.N. (1970). *Visual perception.* San Fransisco, CA: Freeman.

Cowey, A. & Stoerig, P. (1991). The neurobiology of blindsight. *Trends in Neurosciences*, 14, 140–145.

Cowey, A. & Stoerig, P. (1993). Insights into blindsight? *Current Biology*, 3, 236–238.

Cowey, A. & Stoerig, P. (1995). Blindsight in monkeys. *Nature*, 373, 247–249.

Crair, M.C. & Malenka, R.C. (1995). A critical period for long-term potentiation at thalamocritical synapses. *Nature* 375: 325–328.

Crawford, M.L.J., Smith, E.L., Harwerth, R.S. & von Noorden, G.K. (1984). Stereoblind monkeys have few binocular neurons. *Investigative Ophthalmology and Visual Science*, 25, 779–781.

Crawford, M.L.J. & von Noorden, G.K. (1980). Optically induced concomitant strabismus in monkey. *Investigative Ophthalmology and Visual Science*, 19, 1105–1109.

Curcio, C.A., Allen, K.A., Sloan, K.R., Lerea, C.L., Hurley, J.B., Klock, I.B. & Milam, A.H. (1991). Distribution and morphology of human cone photoreceptors stained with anti-blue opsin. *Journal of Comparative Neurology*, 312, 610–624.

Cynader, M. (1983). Prolonged sensitivity to monocular deprivation in dark reared cats: effects of age and visual exposure. *Journal of Neurophysiology*, 43, 1026–1040.

Cynader, M. & Chernenko, G. (1976). Abolition of directional selectivity in the visual cortex of the cat. *Science*, 193, 504–505.

Damasio, H., Grabowski, T., Frank, R., Galaburda, A.M. & Damasio, A.R. (1994). The return of Phineas Gage: clues about the brain from a skull of a famous patient. *Science*, 264, 1102–1105.

Damasio, A.R., Tranel, D. & Damasio, H. (1990). Face agnosia and the neural substrate of memory. *Annual Review of Neuroscience*, 13, 89–109.

Dartnall, H.J.A., Bowmaker, J.K. & Mollon, J.D. (1983). Human visual pigments: microspectrophotometric results from the eyes of seven persons. *Proceedings of the Royal Society of London, Series B*, 220, 115–130.

Davis, G. & Driver, J. (1994). Parallel detection of Kanizsa subjective figures in the human visual system. *Nature*, 371, 791–793.

Daw, N.W., Rader, R.K. Robertson, T.W. & Ariel, M. (1983). Effects of 6–hydroxy-dopamine on visual deprivation in the kitten cortex. *Journal of Neuroscience*, 3, 907–914.

de Jong, B.M., Shipp, S., Skidmore, B., Frackowiak, R.S.J. & Zeki, S. (1994). The cerebral activity related to the visual perception of forward motion in depth. *Brain*, 117, 1039–1054.

De Renzi, E., Perani, D., Carlesimo, G.A., Silveri, M.C. & Fazio, F. (1994). Prosopagnosia can be associated with damage confined to the right hemisphere – an MRI and PET study and a review of the literature. *Neuropsychologia*, 32, 893–902.

De Valois, R.L., Albrecht, D.J. & Thorell, L. (1978). Cortical cells: bar detectors or spatial frequency filters? In *Frontiers in Visual Science*, ed. S.J. Cool & E.L. Smith, Berlin: Springer-Verlag.

De Valois, R.L., Thorell, L.G. & Albrecht, D.G. (1985). Periodicity of striate-cortex-cell receptive fields. *Journal of the Optical Society of America*, 2, 1115–1123.

de Weerd, P., Gattass, R., Desimone, R. & Ungerleider, L.G. (1995) . Responses of cells in monkey visual cortex during perceptual filling-in of an artificial scotoma. *Nature*, 377, 731–734.

Delaye, M. and Tardieu, A. (1983). Short-range order of crystallin proteins accounts for eye lens transparency. *Nature*, 302: 415–417.

Desimone, R., Miller, E.K., Chelazzi, L. & Lueschow, A. (1995). Multiple visual systems in the visual cortex. In *The Cognitive Neurosciences*, ed. M.S. Gazzanida, pp. 475–486. London: MIT Press.

Desimone, R. & Schein, S.J. (1987). Visual properties of neurons in area V4 of the macaque: sensitivity to stimulus form. *Journal of Neurophysiology*, 57, 835–868.

Duffy, C.J. & Wurtz, R.H. (1991). Sensitivity of MST neurons to optic flow stimuli. I.A. continuum of response selectivity to large field stimuli. *Journal of Neurophysiology*, 65, 1329–1345.

Dulai, K.S., Bowmaker, J.K., Mollon, J.D. & Hunt, D.M. (1994). Sequence divergence, polymorphism and evolution of the middle-wave and long-wave visual pigment genes of great apes and old world monkeys. *Vision Research*, 34, 2483–2491.

Eckhorn, R., Bauer, R., Jordan, W., Brosch, M., Kruse, W. Munk, M. & Reitboeck, H.J. (1988). Coherent oscillations: a mechanism for feature linking in the visual cortex. *Biological Cybernetics*, 60, 121–130.

Eckhorn, R., Frien, F., Bauer, R., Woelbern, T. & Kehr, H. (1993). High frequency (60–90 Hz) oscillations in primary visual cortex of awake monkey. *NeuroReport*, 4, 243–246.

Engel, A.K., Konig, P., Gray, C. & Singer, W. (1990). Stimulus-dependent neuronal oscillations in cat visual cortex: intercolumnar interaction as determined by cross-correlation analysis. *European Journal of Neuroscience*, 2, 588–606.

Engel, A.K., Konig, P., Kreiter, A.K., Schillen, T.B. & Singer, W. (1992). Temporal coding in the visual cortex: new vistas on integration in the nervous system. *Trends in Neurosciences*, 15: 218–226.

Engel, A.K., Konig, P. & Singer, W. (1991). Direct physiological evidence for scene segmentation by temporal coding. *Proceedings of the National Academy of Sciences USA*, 88, 9136–9140.

Fagiolini, M., Pizzorusso, T., Berardi, N., Domenici, L. & Maffei, L. (1994). Functional post-natal development of the rat primary visual cortex and the role of visual experience: dark rearing and monocular deprivation. *Vision Research*, 34, 709–720.

Fendrich, R., Wessinger, C.M. & Gazzaniga, M.S. (1992). Residual vision in a scotoma: implications for blindsight. *Science*, 258, 1489–1491.

Ferrera, V.P., Neally, T.A. & Maunsell, J.H.R. (1994). Responses in macaque visual area V4 following the inactivation of the parvocellular and magnocellular LGN pathways. *Journal of Neuroscience*, 14, 2080–2088.

ffytche, D.H., Guy, C. & Zeki, S. (1995). The parallel visual motion inputs into areas V1 and V5 of the human cerebral cortex. *Brain*, 118, 1375–1394.

Fox, K., Sato, H. & Daw, N.W. (1989). The location and function of NMDA receptors in cat and kitten visual cortex. *Journal of Neuroscience*, 9, 2443–2454.

Freeman, R.D. & Pettigrew, J.D. (1973). Alterations of visual cortex from environmental asymmetries. *Nature*, 246, 359–360.

Fujita, I, Tanaka, K., Ito, M. & Cheng, K. (1992). Columns for visual features of objects in monkey inferotemporal cortex. *Nature*, 360, 343–346.

Ghose, G.M. & Freeman, R.D. (1992). Oscillatory discharge in the visual system: does it have a functional role? *Journal of Neurophysiology*, 68, 1558–1574.

Girard, P. & Bullier, J. (1988). Visual activity in area V2 during reversible inactivation of area 17 in the macaque monkey. *Journal of Neurophysiology*, 62, 1287–1302.

Girad, P., Salin, P.A. and Bullier, J. (1992). Response selectivity of neurons in area MT of the macaque monkey during reversible activation of V1. *Journal of Neurophysiology*, 67, 1437–1446.

Goodale, M.A. & Milner, A.D. (1992). Separate visual pathways for perception and action. *Trends in Neuroscience*, 15, 20–25.

Goodale, M.A., Milner, A.D., Jacobsen, L.S. & Carey, D.P. (1991). A neurological dissociation between perceiving objects and grasping them. *Nature*, 349, 154–156.

Gray, C.M., Engel, A.K., Konig, P. & Singer, W. (1990). Stimulus-dependent neuronal oscillations in cat visual cortex: receptive field properties and feature dependence. *European Journal of Neuroscience*, 2, 607–619.

Gray, C.M., Engel, A.K., Konig, P. & Singer, W. (1992). Synchronization of oscillatory neuronal responses in cat striate cortex: temporal properties. *Visual Neuroscience*, 8, 337–347.

Gray, C.M., Konig, P., Engel, A.K. & Singer, W. (1989). Oscillatory responses in cat visual cortex exhibit inter-columnar synchronization which reflects global stimulus properties. *Nature*, 338, 334–337.

Gray, C.M. & Singer, W. (1989). Stimulus specific neuronal oscillations in the orientation columns of cat visual cortex. *Proceedings of the National Academy of Sciences USA*, 86, 1698–1702.

Graziano, M.S.A., Andersen, R.A. & Snowden, R.J. (1994). Tuning of MST neurons to spiral motions. *Journal of Neuroscience*, 14, 54–67.

Gregory, R.L. (1972). Cognitive contours. *Nature*, 238, 51–52.

Grosof, D.H., Shapley, R.M. & Hawken, M.J. (1993). Macaque V1 neurons can signal 'illusory' contours. *Nature*, 365, 550–552.

Grusser, O.J. & Landis, T. (1991). *Visual Agnosias*. London: MacMillan Press.

Gulyas, B. & Roland, P.E. (1991). Cortical fields participating in form and colour discrimination in the human brain. *Neuroreport*, 2, 585–588.

Halligan, P.W. & Marshall, J.C. (1991). Left neglect for near but not far space in man. *Nature*, 350, 498–500.

Hamann, S.B., Stefancci, L., Squire, L.R., Adolphs, R., Tranel, D., Damasio, H. & Damasio, A. (1996). Recognising facial emotion. *Nature*, 379, 497.

Hammond, P., Movat, G.S.V. & Smith, A.T. (1985). Motion after-effects in cat striate cortex elicited by moving gratings. *Experimental Brain Research*, 60, 411–416.

Harl, R., Salmelin, R. & Tissarl, S.O. (1994). Visual stability during eyeblinks. *Nature*, 367, 121–122.

Harries, M.H. & Perrett, D.I. (1991). Modular organization of face processing in temporal cortex: physiological evidence and possible anatomical correlates. *Journal of Cognitive Neuroscience*, 3, 9–24.

Hasselmo, M.E., Rolls, E.T. & Baylis, G.C. (1989). The role of expression and identity in the face-selective responses of neurons in the temporal visual cortex of the monkey. *Experimental Brain Research,* 32, 203–218.

Haxby, J.V., Grady, C.L., Horwitz, B., Ungerleider, L.G., Mishkin, M., Carson, R.E., Herscovitch, P., Schapiro, M.B., Rapoport, S.I. (1991). Dissociation of spatial and object visual processing pathways in human extrastriate cortex. *Proceedings of the National Academy of Sciences USA,* 88, 1621–1625.

Hebb, D.O. (1949). *The Organization of Behavior.* New York: Wiley.

Hendry, S.H.C. & Yoshioka, T. (1994). A neurochemically distinct third channel in the macaque dorsal lateral geniculate nucleus. *Science* , 264, 575–577.

Heywood, C.A., Gadotti, A. & Cowey, A. (1992). Cortical area V4 and its role in the perception of color. *Journal of Neuroscience,* 12, 4056–4065.

Hirsch, H.V.B. & Spinelli, D.N. (1970). Visual experience modifies distribution of horizontally and vertically orientated receptive fields in cats. *Science,* 168, 869–871.

Hubel, D.H. (1989). *Eye, Brain and Vision.* New York: Scientific American Library.

Hubel, D.H. & Wiesel, T.N. (1962). Receptive fields, binocular interaction and functional architecture in the cat's visual cortex. *Journal of Physiology,* 160, 106–154.

Hubel, D.H. & Wiesel, T.N. (1965). Binocular interaction in striate cortex of kittens reared with artificial squint. *Journal of Neurophysiology,* 28, 1041–1059

Hubel, D.H. & Wiesel, T.N. (1970) The period of susceptibility to the physiological effects of unilateral lid closure in kittens. *Journal of Physiology,* 206. 419–436.

Hubel, D.H. & Wiesel, T.N. (1977). Functional architecture of macaque monkey visual cortex (Ferrier Lecture). *Proceedings of the Royal Society, London, series B,* 198, 1–59.

Hung, I.F., Crawford, M.C.J. & Smith, E.L. (1995). Spectacle lenses alter eye growth and the refractive status of young monkeys. *Nature Medicine,* 1, 761–765.

Hunt, D.M., Dulai. K.S., Bowmaker, J.K., Mollon, J.D. (1995). The chemistry of John Dalton's color blindness. *Science,* 267, 984–987.

Hurley, J.B., Dizhoor, A.M., Ray, S. & Stryer, L. (1993). Recoverin's role: conclusion withdrawn. *Science,* 260, 740.

Ibbotson, R.E., Hunt, D.M., Bowmaker, J.K. & Mollon, J.D. (1992). Sequence divergence and copy number of the middle- and long-wave photopigment genes in Old World monkeys. *Proceedings of the Royal Society of London, Series B,* 247, 145–154.

Jakobsen, L.S., Archibald, Y.M., Carey, D.P. & Goodale, M.A. (1991). A kinematic analysis of reaching and grasping movements in a patient recovering from optic ataxia. *Neuropsychologia,* 29, 803–809.

Jordan, G. & Mollon, J.D. (1993). A study of women heterozygous for colour deficiencies. *Vision Research,* 33, 1495–1508.

Kaas, J.H. (1989). Why does the brain have so many visual areas? *Journal of Cognitive Neuroscience,* 1, 121–135.

Kaas, J.H. (1992). Do humans see what monkeys see? *Trends in Neurosciences,* 15, 1–3.

Kandel, E.R. & Schwartz, J.H. (1982). *Principles of Neural Science*. New York: Elsevier Science.

Kaufman, L. (1974). *Sight and Mind*. Oxford: Oxford University Press.

Kawamura, S. (1993). Rhodopsin phosphorylation as a mechanism of cyclic GMP phosphodiesterase regulation by S-modulin. *Nature*, 362, 855–857.

Kirkwood, A. & Bear, M.F. (1994). Homosynaptic long-term depression in the visual cortex. *Journal of Neuroscience*, 14, 3403–3412.

Kirkwood, A., Lee, H.-K. & Bear, M.F. (1995). Co-regulation of long-term potentiation and experience-dependent synaptic plasticity in visual cortex by age and experience. *Nature*, 375, 328–331.

Konorski, J. (1967). Integrative activity of the brain: an interdisciplinary approach. Chicago: University of Chicago Press.

Koretz, J.F. & Handelman, G.H. (1988). How the human eye focuses. *Scientific American*, 259, 92–99.

Kosslyn, S.M. & Oschner, K.N. (1994). In search of occipital activation during visual mental imagery. *Trends in Neurosciences*, 17, 290–292.

Koutalos, Y. & Yau, K.-W. (1993). A rich complexity emerges in phototransduction. *Current Opinion in Neurobiology*, 3, 513–519.

Kreiter, A.K. & Singer, W. (1992). Oscillatory neuronal responses in the visual cortex of the awake macaque monkey. *European Journal of Neuroscience*, 4, 369–375.

Land, E.H. (1964). The retinex. *Scientific American*, 52, 247–264.

Land, E.H. (1983). Recent advances in retinex theory and some implications for cortical computations. *Proceedings of the National Academy of Sciences USA*, 80, 5163–5169.

Land, E.H., Hubel, D., Livingstone, M., Perry, S. & Burns, M. (1983). Color-generating interactions across the corpus callosum. *Nature*, 303, 616–618.

Landis, T., Regard, M., Bliestle, A. & Kleihues, P. (1988). Prosopagnosia and agnosia for noncanonical views. *Brain*, 111, 1287–1297.

Le Bihan, D., Turner, R., Zeffino, T.A., Cuenod, C.A., Jezzard, P. & Bonnerd, V. (1993). Activation of human primary visual-cortex during visual recall: a magnetic-resonance-imaging study. *Proceedings of the National Academy of Sciences USA*, 90, 11802–11805.

Lennie, P. & D'Zmura, M. (1988). Mechanisms of color vision. *CRC Critical Reviews in Neurobiology*, 3, 333–400.

LeVay, S., Stryker, M.P. & Shatz, C.J. (1978). Ocular dominance columns and their development of layer IV of the cat's visual cortex: a quantitative study. *Journal of Comparative Neurology*, 179, 223–244.

Livingstone, M.S. (1991). Visually evoked oscillations in monkey striate cortex. *Society for Neuroscience (Abstracts)*, 17, 73.3.

Livingstone, M. & Hubel, D. (1988). Segregation of form, color, movement, and depth: anatomy, physiology and perception. *Science*, 240, 740–749.

Lømo, T. (1966). Frequency potentiation of excitatory synaptic activity in the dentate area of the hippocampal formation. *Acta Physiologica Scandinavic,* 68 (Suppl), 277.

Lowel, S. & Singer, W. (1993). Monocularly induced 2–deoxyglucose patterns in visual cortex and lateral geniculate nucleus of the cat: II. Awake animals and strabismic animals. *European Journal of Neuroscience,* 5, 857–869.

Marr, D. (1982). *Vision: A Computational Investigation into the Human Representation and Processing of Visual Information.* San Fransisco: W H Freeman.

Marshall, J.C. & Halligan, P.W. (1988). Blindsight and insight in visuo-spatial neglect. *Nature,* 336, 766–767.

Mather, G. & Moulden, B. (1980). A simultaneous shift in apparent direction: further evidence for a 'distribution-shift' model of direction coding. *Quarterly Journal of Experimental Psychology,* 32, 325–333.

Matin, L., Picoult, E., Stevens, J., Edwards, M. & MacArthur, R. (1982). Oculoparalytic illusion: visual-field dependent spatial mislocations by humans partially paralysed by curare. *Science,* 216, 198–201.

Maunsell, J.H.R., Nealey, T.A. & DePriest, D.D. (1990). Magnocellular and parvocellular contributions to responses in the middle temporal visual area (MT) of the macaque monkey. *Journal of Neuroscience,* 10, 3323–3334.

Maunsell, J.H.R. & Newsome, W.T. (1987). Visual processing in monkey extrastriate cortex. *Annual Review of Neuroscience,* 10, 363–401.

Merbs, S.L. & Nathans, J. (1992a). Absorption spectra of the hybrid pigments responsible for anomalous colour vision. *Science,* 258, 464–466.

Merbs, S.L. & Nathans, J. (1992b). Absorption spectra of human cone pigments. *Nature* 356, 433–435.

Miller, E.K. & Desimone, R. (1994). Parallel neuronal mechanisms for short-term memory. *Science,* 263, 520–522.

Miller, E.K., Gross, P.M. & Gross, C.G. (1991). Habituation-like decrease in the responses of neurons in the inferior temporal cortex of the macaque. *Visual Neuroscience,* 7, 357–362.

Mishkin, M., Ungerleider, L.G. & Macko, K.A. (1983). Object vision and spatial vision: two cortical pathways. *Trends in Neurosciences,* 6, 414–417.

Miyashita, Y. & Chang, H.S. (1988). Neuronal correlate of pictorial short-term memory in the primate temporal cortex. *Nature,* 331, 68–70.

Mohler, C.W. & Wurtz, R.H. (1977). The role of striate cortex and superior colliculus in visual guidance of saccadic eye movements in monkeys. *Journal of Neurophysiology,* 40, 74–94.

Mollon, J.D. (1989). 'Tho' she kneel'd in that place where they grew . . .' The uses and origins of primate colour vision. *Journal of Experimental Biology,* 146, 21–38.

Mollon, J.D. (1991). Edwin Herbert Land (1909–1991). *Nature,* 350, 379–380.

Mollon, J.D. & Bowmaker, J.K. (1992). The spatial arrangement of cones in the primate fovea. *Nature,* 360, 677–679.

Moran, J. and Desimone, R. (1985). Selective attention gates visual processing in the extrastriate visual cortex. *Science,* 229, 782–784.

Morgan, M.J. (1994). When it pays not to see. *Nature,* 371, 473.

Morgan, M.J., Adam, A. & Mollon, J.D. (1992). Dichromates detect colour-camouflaged objects that are not detected by trichromates. *Proceedings of the Royal Society of London, Series B,* 248, 291–295.

Moscovitch, M., Behrmann, M. & Winocur, G. (1994). Do PETS have long or short ears? Mental imagery and neuroimaging. *Trends in Neurosciences,* 17, 292–294.

Movshon, J.A., Adelson, E.H., Gizzi, M.S. & Newsome, W.T. (1985). The analysis of moving visual patterns. In *Pattern Recognition Mechanisms,* ed. C. Chagas, R. Gattass & C. Gross, pp. 117–151. Vatican City: Pontifical Academy of Sciences.

Muir, D.W. & Mitchell, D.E. (1975). Behavioural defects in cats following early selected visual exposure to contours of a single orientation. *Brain Research,* 85, 459–477.

Murasagi, C.M., Salzman, C.D. & Newsome, W.T. (1993). Microstimulation in visual area MT: effects of varying pulse amplitude and frequency. *Journal of Neuroscience,* 13, 1719–1729.

Nagy, A.L. & Wolf, S. (1993). Red–green discrimination in peripheral-vision. *Vision Research,* 33, 235–242.

Nahm, F.K.D., Tranel, D., Damasio, H. & Damasio, A.R. (1993). Cross-modal associations and the human amygdala. *Neuropsychologia,* 31, 727–744.

Nakamura, K., Mikami, A. & Kubota, K. (1992). Oscillatory activity related to short-term visual memory in monkey temporal pole. *NeuroReport* , 3, 17–120.

Nathans, J., Davenport, C.M., Maumenee, I.H., Lewis, R.A., Hejtmark, J.F., Litt, M., Lovrien, E., Weleber, R., Bachynski, B., Zwas, F., Klingaman, R. & Fishman, G. (1989). Molecular genetics of human blue cone monochromacy. *Science,* 245, 831–838.

Nathans, J., Piantanida, T.P., Eddy, R.L., Shows, T.B. & Hogness, D.S. (1986a). Molecular genetics of inherited variation in human colour vision. *Science,* 232, 203–222.

Nathans, J., Thomas, D. & Hogness, D.S.(1986b). Molecular genetics of human color vision: the genes encoding blue, green and red pigments. *Science,* 232, 193–202.

Nealey, T.A. & Maunsell, J.H.R. (1994). Magnocellular and parvocellular contributions to the responses of neurons in the macaque striate cortex. *Journal of Neuroscience,* 14, 2080–2088.

Neitz, J. & Jacobs, G.H. (1986). Polymorphism of the long-wavelength cone in normal human colour vision. *Nature* 323, 623–625.

Neitz, J. & Neitz, M. (1993). All common red–green color defects can be explained as arising by the same simple mechanism. *Investigative Ophthalmology and Visual Sciences,* 34, 911.

Neitz, M. & Neitz, J. (1995). Numbers and ratios of visual pigment genes for normal red–green color vision. *Science,* 267, 1013–1016.

Neitz, M., Neitz, J. & Jacobs, G.H. (1991). Spectral tuning of pigments underlying red–green color vision. *Science*, 252, 971–974.

Neitz, J., Neitz, M. & Jacobs, G.H. (1993). More than three different cone pigments among people with normal color vision. *Vision Research,* 33, 117–122.

Newsome, W.T. & Paré, E.B. (1988). A selective impairment of motion perception following lesions of the middle temporal visual area (MT). *Journal of Neuroscience*, 8, 2201–2211.

Nicholls, J.G., Martin, A.R. & Wallace, B.G. (1992). *From Neuron To Brain*, 3rd edn. Cambridge, MA: Sinauer.

Normann, R.A. & Perlman, I. (1979). The effects of background illumination on the photoresponses of red and green cones. *Journal of Physiology*, 286, 491.

Olson, C.R. & Freeman, R.D. (1975). Progressive changes in kitten striate cortex during monocular vision. *Journal of Neurophysiology,* 38, 26–32.

Olson, C.R. & Freeman, R.D. (1980). Profile of the sensitive period for monocular deprivation in kittens. *Experimental Brain Research*, 39, 17–21.

Orban, G.A.., Lagae, L., Verri, A., Raiguel, S., Xiao, D., Maes, H. & Torre, V. (1992). First-order analysis of optical flow in monkey brain. *Proceedings of the National Academy of Sciences USA*, 89, 2595–2599.

O'Shea, R.P., Blackburn, S.G. & Ono, H. (1993). Aerial perspective, contrast, and depth perception. *Investigative Ophthalmology and Visual Science*, 34, 1185.

Ott, M. & Schaeffel, F. (1995). A negatively powered lens in the chameleon. *Nature,* 373, 692–694.

Pasik, P. & Pasik, T. (1982). Visual functions in monkeys after total removal of visual cerebral cortex. In *Contributions to Visual Physiology,* vol. 7, ed. W.D. Neff, pp. 147–200. New York: Academic Press.

Peng, Y.W., Robishaw, J.D., Levine, M.A. & Yau, K.W. (1992). Retinal rods and cones have distinct G protein β and γ subunits. *Proceedings of the National Academy of Sciences USA,* 89, 10882–10886.

Perrett, D.I., Hietnan, J.K., Oram, M.W. & Benson, P.J. (1992). Organisation and functions of cells responsive to faces in the temporal cortex. *Philosophical Transactions of the Royal Society of London, Series B.* 335, 23–30.

Perrett, D.I. & Oram, M.W. (1993). Neurophysiology of shape processing. *Image and Vision Computing,* 11, 317–333.

Perrett, D.I., Rolls, E.T. & Caan, W. (1982). Visual neurones responsive to faces in the monkey temporal cortex. *Experimental Brain Research*, 47, 329–342.

Perry, V.H. & Cowey, A. (1985). The ganglion cell and cone distributions in the monkey's retina: implications for central magnification factors. *Vision Research*, 25, 1795–1810.

Peterhans, E. & von der Heydt, R. (1989). Mechanisms of contour perception in monkey visual cortex: I. Contours bridging gaps. *Journal of Neuroscience*, 9, 1749–1763.

Peterhans, E. & von der Heydt, R. (1991). Subjective contours – bridging the gap between psychophysics and physiology. *Trends in Neurosciences,* 14, 112–119.

Pettigrew, J.D. (1974). The effect of visual experience on the development of stimulus specificity by kitten cortical neurons. *Journal of Physiology*, 237, 49–74.

Pettigrew, J.D. & Freeman, R.D. (1973). Visual experience without lines: effects on developing cortical neurons. *Science*, 182, 599–601.

Poggio, G.F. & Fischer, B. (1977). Binocular interaction and depth sensitivity in striate cortex. and prestriate cortex of behaving rhesus monkey. *Journal of Neurophysiology*, 40, 1392–1405.

Poggio, G.F. & Poggio, T. (1984). The analysis of stereopsis. *Annual Review of Neuroscience*, 7, 379–412.

Posner, M.I. & Petersen, S.E. (1990). The attention system of the human brain. *Annual Review of Neuroscience*, 13, 25–42.

Puce, A. Allison, T., Gore, J.C. & McCarthy, G. (1995). Face-sensitive regions in human extrastriate cortex studied by functional MRI. *Journal of Neurophysiology*, in press.

Ramachandran, V.S. (1988). Perception of shape from shading. *Nature*, 331, 163–166.

Ramachandran, V.S. (1992). Blind spots. *Scientific American*, 266, 86–91.

Rao, R.V., Cohen, G.B. & Oprian, D.D. (1994). Rhodopsin mutation G90D and a molecular mechanism for congenital nightblindness. *Nature*, 367, 639–642.

Raviola, E. & Wiesel, T.N. (1985). An animal model of myopia. *New England Journal of Medicine*, 312, 1609–1615.

Reiter, H.O. & Stryker, M.P. (1988). Neural plasticity without postsynaptic action potentials: less active inputs become dominant when kitten visual cortex cells are pharmacologically inhibited. *Proceedings of the National Academy of Sciences, USA*, 85, 3623–3627.

Rodiek, R.W. (1988). The Primate retina. In *Comparative Primate Biology*, Vol. 4, *Neurosciences*. ed. Steklis, H.D. & Erwin, J. pp. 203–278. New York: A.R. Liss.

Rodman, H.R., Gross, C.G. and Albright, T.D. (1989a). Afferent basis of visual response properties in area MT of the macaque. I. Effects of striate cortex removal. *Journal of Neuroscience*, 9, 2033–2050.

Rodman, H.R., Gross, C.G. and Albright, T.D. (1989b). Afferent basis of visual response properties in area MT of the macaque. II. Effects of superior colliculus removal. *Journal of Neuroscience*, 10, 2033–2050.

Roe, A.W. & Ts'o, D. (1995). Visual topography in primate V2: multiple representation across functional stripes. *Journal of Neuroscience*, 15, 3689–3715.

Roland, P.E. & Gulyas, B. (1994). Visual imagery and visual representation. *Trends in Neurosciences*, 17, 281–287.

Rolls, E.T. & Baylis, G.C. (1986). Size and contrast have only small effects on the responses to faces of neurons in the cortex of the superior temporal sulcus of the monkey. *Experimental Brain Research*, 65, 38–48.

Rolls, E.T., Baylis, G.C., Hasselmo, M.E. & Nalwa, V. (1989). The effect of learning on the face selective responses of neurons in the cortex in the superior temporal sulcus of the monkey. *Experimental Brain Research*, 76, 153–164.

Rolls, E.T. & Tovée, M.J. (1994). Processing speed in the cerebral cortex, and the neurophysiology of backward masking. *Proceedings of the Royal Society of London, Series B,* 257, 9–15.

Rolls, E.T. & Tovée, M.J. (1995). Sparseness of the neuronal representation of stimuli in the primate temporal visual cortex. *Journal of Neurophysiology,* 73, 713–726.

Rolls, E.T., Tovée, M.J., Purcell, D.G., Stewart, A.L. & Azzopardi, P. (1994). The responses of neurons in the temporal cortex of primates and face identification and detection. *Experimental Brain Research,* 101, 473–484.

Rolls, E.T., Tovée, M.J. & Ramachandran, V.S. (1993). Visual learning reflected in the responses of neurons in the temporal visual cortex of the macaque. *Society for Neuroscience Abstracts,* 19, 27.

Rushton, W.A. (1961). Rhodopsin measurement and dark adaptation in a subject deficient in cone vision. *Journal of Physiology,* 156, 193–205.

Saito, H.-A., Yuki, M., Tanaka, K., Hikosaka, K., Fukada, Y. & Iwai, E. (1986). Integration of direction signals of image motion in the superior temporal sulcus of the macaque. *Journal of Neuroscience,* 6, 145–157.

Salzman, C.D. & Newsome, W.T. (1994). Neural mechanisms for forming a perceptual decision. *Science,* 264, 231–237.

Sary, G., Vogels, R. & Orban, G.A. (1993). Cue-invariant shape selectivity of macaque inferior temporal neurons. *Science,* 260, 995–997.

Saund, E. (1992). Putting knowledge into visual shape representation. *Artificial Intelligence,* 54, 71–119.

Scannell, J.W., Blakemore, C. & Young, M.P. (1995). Analysis of connectivity in the cat cerebral cortex. *Journal of Neuroscience,* 15, 1463–1483.

Schaeffel, F., Glasser, A. & Howland, H.C. (1988). Accommodation, refractive error and eye growth in the chicken. *Vision Research,* 28, 639–657.

Schaeffel, F., Troilo, D., Wallman, J. & Howland, H.C. (1990). Developing eyes that lack accommodation grow to compensate for imposed defocus. *Visual Neuroscience,* 4, 177–183.

Schiller, P.H. & Lee, K. (1991) The role of primate extrastriate area V4 in vision. *Science,* 251, 1251–1253.

Schiller, P.H. & Malpeli, J.G. (1978). Functional specifity of lateral geniculate nucleus laminae of the rhesus monkey. *Journal of Neurophysiology,* 41, 788–797.

Schlagger, B.L., Fox, K. & O'Leary, D.D.M. (1993). Postsynaptic control of plasticity in developing somatosensory cortex. *Nature,* 364, 623–626.

Sekuler, R. & Blake R. (1994). *Perception* (3rd Ed.). New York: McGraw-Hill Inc.

Sereno, M.I., Dale, A.M., Reppas, J.B., Kwong, K.K., Belliveau, J.W., Brady, T.J., Rosen, B.R. & Tootell, R.B.H. (1995). Borders of multiple visual areas in humans revealed by functional magnetic resonance imaging. *Science,* 268, 889–893.

Sergent, J., Ohta, S. & MacDonald, B. (1992). Functional neuroanatomy of face and object processing: a positron emission tomography study. *Brain,* 115, 15–36.

Sergent, J. & Signoret, J.L. (1992). Varieties of functional deficits in prospagnosia. *Cerebral Cortex*, 2, 375–388.

Shaw, C. & Cynader, M. (1984). Disruption of cortical activity prevents alterations of ocular dominance in monocularly deprived kittens. *Nature*, 308, 731–734.

Sherk, H. & Stryker, M.P. (1976). Quantitative study of cortical orientation selectivity in visually inexperienced kitten. *Journal of Neurophysiology*, 39, 63–70.

Shipp, S., de Jong, B.M., Zihl, J., Frackowiak, R.S.J. & Zeki, S. (1994). The brain activity related to residual activity in a patient with bilateral lesions of V5. *Brain*, 117, 1023–1038.

Shipp, S. & Zeki, S. (1985). Segregation of pathways leading from area V2 to areas V4 and V5 of the macaque monkey visual cortex. *Nature*, 315, 322–325.

Sillito, A.M., Jones, E.H., Gerstein, G.L. & West, D.C. (1994). Feature-linked synchronisation of thalamic relay cell firing induced by feedback from the visual cortex. *Nature*, 369, 479–482.

Singer, W. (1993). Synchronization of cortical activity and its putative role in information processing and learning. *Annual Review of Physiology*, 55, 349–374.

Singer, W. (1994). A new job for the thalamus. *Nature*, 369, 444–445.

Stein, J.F. (1992). The representation of egocentric space in the posterior parietal cortex. *Behavioural and Brain Sciences*, 15, 691–700.

Stevens, J.K., Emerson, R.C., Gerstein, G.L., Kallos, T., Neufeld, G.R., Nichols, C.W. & Rosenquist, A.C. (1976). Paralysis of the awake human. Visual perceptions. *Vision Research*, 16, 93–98.

Stevens, K.A. (1983). Evidence relating subjective contours and interpretations involving interposition. *Perception*, 12, 491–500.

Stone, R.A., Laties, A.M., Raviola, E. & Wiesel, T.N. (1988). Increase in retinal vasoactive intestinal polypeptide after eyelid fusion in primates. *Proceedings of the National Academy of Sciences USA*, 86, 704–706.

Stonor, G.R. & Albright, T.D. (1993). Image segmentation cues in motion processing: implications for modularity in vision. *Journal of Cognitive Neuroscience*, 5, 129–149.

Stryker, M.P. (1992). Elements of visual perception. *Nature*, 360, 301–302.

Stryker, M.P. & Strickland, S.L. (1984). Physiological segregation of ocular dominance columns depends on the pattern of afferent electrical activity. *Investigative Ophthalmology and Visual Science*, 25 (Suppl), 278.

Sung, C.H., Schneider, B.G., Agarwal, N., Papermaster, D.S. & Nathans, J. (1991). Functional heterogeneity of mutant rhodopsins responsible for autosomal dominant retinitis pigmentosa. *Proceedings of the National Academy of Sciences USA*, 88, 8840–8844.

Tanaka, K. (1992). Inferotemporal cortex and higher visual functions. *Current Opinion in Neurobiology*, 2, 502–505.

Tanaka, D.W., Ogawa, S. & Ugurbil, K. (1992). Mapping the brain with MRI. *Current Biology*, 2, 525–528.

Tanaka, K., Saito, H., Fukada, Y. & Moriya, M. (1991). Coding visual images of objects in the inferotemporal cortex of the macaque monkey. *Journal of Neurophysiology*, 66, 170–189.

Tootell, R.B.H., Reppas, J.B., Dale, A.M., Look, R.B., Sereno, M.I., Malach, R., Brady, T.J. & Rosen, B.R. (1995a). Visual motion aftereffect in human cortical area MT revealed by functional magnetic resonance imaging. *Nature,* 375, 139–141.

Tootell, R.B.H., Reppas, J.B., Kwong, K.K., Malach, R., Born, R.T., Brady, T.J., Rosen, B.R. & Belliveau, J.W. (1995b). Functional analysis of human MT and related cortical visual cortical areas using magnetic resonance imaging. *Journal of Neuroscience*, 15, 3215–3230.

Tovée, M.J. (1993). Colour vision in New World monkeys and the single-locus X-chromosome theory. *Brain, Behaviour and Evolution,* 42, 116–127.

Tovée, M.J. (1994a). The molecular genetics and evolution of primate colour vision. *Trends in Neurosciences*, 17, 30–37.

Tovée, M.J. (1994b). How fast is the speed of thought? *Current Biology,* 4, 1125–1127.

Tovée, M.J. (1995a). Ultra-violet photoreceptors in the animal kingdom: their distribution and function. *Trends in Ecology and Evolution*, 11, 455–460.

Tovée, M.J. (1995b) Face recognition: what are faces for? *Current Biology*, 5, 480–482.

Tovée, M.J., Bowmaker, J.K. & Mollon, J.D. (1992). The relationship between cone pigments and behavioural sensitivity in a New World monkey (*Callithrix jacchus jacchus*). *Vision Research*, 32, 867–878.

Tovee, M.J. & Cohen-Tovee, E.M. (1993). The neural substrates of face processing models: a review. *Cognitive Neuropsychology*, 10, 505–528.

Tovée, M.J. & Rolls, E.T. (1992). Oscillatory activity is not evident in the primate temporal visual cortex with static stimuli. *NeuroReport,* 3, 369–372.

Tovée, M.J. & Rolls, E.T. (1995). Information encoded in short firing rate epochs by single neurons in the primate temporal visual cortex. *Visual Cognition,* 2, 35–58.

Tovée, M.J., Rolls, E.T. & Azzopardi, A. (1994). Translation invariance in the responses to faces of single neurons in the temporal visual cortical areas of the alert macaque. *Journal of Neurophysiology*, 72, 1049–1060.

Tovée, M.J., Rolls, E.T., Treves, A. & Bellis, R.P. (1993). Information encoding and the responses of single neurons in the primate temporal visual cortex. *Journal of Neurophysiology*, 70, 640–654.

Tranel, D. & Hyman, B.T. (1990). Neuropsychological correlates of bilateral amygdala damage. *Archives of Neurology*, 47, 349–355.

Ts'o, D.Y., Frostig, R.D., Lieke, E.E. & Grinvald, A. (1990). Functional organization of primate visual cortex revealed by high-resolution optical imaging. *Science*, 249, 417–420.

Ts'o, D.Y. & Gilbert, C.D. (1988). The organization of chromatic and spatial interactions in the primate striate cortex. *Journal of Neuroscience*, 8, 1712–1727.

Ts'o, D.Y. & Roe, A.W. (1995). Functional compartments in visual cortex: segregation and interaction. In *The Cognitive Neurosciences*, ed. M.S. Gazzaniga, pp. 325–337. Cambridge, MA: MIT Press.

Uchikawa, K. & Sato, M. (1995). Saccadic suppression of achromatic and chromatic responses measured by increment-threshold spectral sensitivity. *Journal of The Optical Society of America*, 12, 661–666.

Underwood, N.R. & McConkie, G.W. (1985). Perceptual span for letter distinctions during reading. *Reading Research Quarterly,* 20, 1156–1162.

Ungerleider, L.G. & Haxby, J.V. (1994). 'What' and 'where' in the human brain. *Current Opinion in Neurobiology*, 4, 157–165.

van den Berg, T.J.T.P. & Tan, K.E.W.P. (1994). Light transmittance of the human cornea from 320 to 700 nm for different ages. *Vision Resion,* 34, 1453–1456.

Van Essen, D.C., Anderson, C.H. & Felleman, D.J. (1992). Information processing in the primate visual system: an integrated systems perspective. *Science*, 255: 419–423.

Van Essen, D.C. & Deyoe, E.A. (1995). Concurrent processing in the primate visual cortex. In *The Cognitive Neurosciences*. ed. M.S. Grazzanida, pp. 384–400. London: MIT Press.

Van Essen, D.C. & Gallant, J.L. (1994). Neural mechanisms of form and motion processing in the primate visual system. *Neuron*, 13, 1–10.

van Norren, D. and Schelkens, P. (1990). Blue light hazard in the rat. *Vision Resion,* 30, 1517–1520.

Van Sluyters, R.C. & Blakemore, C. (1973). Experimental creation of unusual properties in visual cortex of kittens. *Nature,* 246, 505–508.

Volkman, F., Riggs, L. & Moore, R. (1980) Eyeblinks and visual suppression. *Science*, 207, 900–902.

von Cramon, D. & Kerkhof, G. (1993). On the cerebral organization of elementary visuospatial perception. In *Functional Organization of the Human Visual Cortex,* ed. B. Gulyas, D. Ottoson & P.E. Roland, pp. 211–231. Oxford: Pergamon Press.

von der Heydt, R., Peterhans, E. & Dürstler, M.R. (1992). Periodic-pattern-selective cells in monkey striate cortex. *Journal of Neuroscience*, 12, 1416–1434.

von der Malsburg, C. (1981). *The Correlation Theory of Brain Function* (Internal Report 81–2). Munich: Max-Plank-Institute for Biophysical Chemistry.

Wallach, H. & O'Connell, D.N. (1953). The kinetic depth effect. *Journal of Experimental Psychology*, 45, 205–217.

Wallman, J. (1994). Nature and nurture of myopia. *Nature*, 371, 201–202.

Wang, G., Tanaka, K. & Tanifuji, M. (1994). Optical imaging of functional organization in macaque inferotemporal cortex. *Society for Neuroscience Abstracts*, 20, 316.

Watson, J.D.G., Frackowiak, R.S.J. & Zeki, S. (1993a). Functional separation of colour and motion centres in human visual cortex. In *Functional Organization of the Human Visual Cortex*. ed. B. Gulyas, D. Ottoson & P.E. Roland. pp. 317–329. Oxford: Pergamon Press.

Watson, J.D., Myers, R., Frackowiak, R.S.J., Hajnal, J.V., Woods, R.P., Mazziota, J.C., Shipp, S. & Zeki, S. (1993b). Area V5 of the human brain: evidence from a combined study using positron emmission tomography and and magnetic resonance imaging. *Cerebral Cortex,* 3, 79–94.

Weitz, C.J., Miyake, Y., Shinzato, K., Montag, E., Zrenner, E., Went, L.N. & Nathans, J. (1992). Human tritanopia associated with two amino acid substitutions in the blue-sensitive opsin. *American Journal of Human Genetics,* 50, 498–507.

Weller, R.E. (1988). Two cortical visual systems in Old World and New World primates. In *Progress in Brain Research,* vol. 75, ed. T.P. Hicks & G. Benedek, pp. 293–306. Amsterdam: Elsevier.

White, H.E. & Levatin, P. (1962). 'Floaters' in the eye. *Scientific American,* 206, 119–127.

Wiesel, T.N. & Hubel, D.H. (1963). The effects of visual deprivation on the morphology and physiology of cell's lateral geniculate body. *Journal of Neurophysiology,* 26, 978–993.

Wiesel, T.N. & Hubel, D.H. (1965). Comparisons of the effects of unilateral and bilateral eye closure on single unit responses in kittens. *Journal of Neurophysiology,* 28, 1029–1040.

Wiesel, T.N. & Raviola, E. (1977). Myopia and eye enlargement after neonatal lid fusion in monkeys. *Nature,* 266, 66–68.

Wiesenfield, K. & Moss, F. (1995). Stochastic resonance and the benefits of noise: from ice ages to crayfish and squids. *Nature,* 375, 33–36.

Wild, H.M., Butler, S.R., Carden, D. & Kulikowski, J.J. (1985). Primate cortical area V4 is important for colour constancy but not wavelength discrimination. *Nature,* 313, 133–135.

Wildsoet, C.F. & Wallman, J. (1992). Optic nerve section affects ocular compensation for spectacle lenses. *Investigative Ophthalmology and Visual Science,* 33 *(Suppl.),* 1053.

Williams, K., Russel, S.L., Shen, Y.M. & Molinoff, P.B. (1993). Developmental switch in the expression of NMDA receptors occurs in vivo and in vitro. *Neuron,* 10, 267–278.

Wilson, F.A.W., O'Scalaidhe, S.P. & Goldman-Rakic, P.S. (1993). Dissociation of object and spatial processing domains in primate prefrontal cortex. *Science,* 260, 1955–1958.

Winderickx, J., Battisti, L., Motulsky, A.G. & Deeb, S.S. (1992a). Selective expression of human X-chromosome linked green opsin genes. *Proceedings of the National Academy of Sciences USA,* 89, 9710–9714.

Winderickx, J., Lindsey, D.T., Sanocki, E., Teller, D.Y., Motulsky, A.G. & Deeb, S.S. (1992b). Polymorphism in red photopigment underlies variation in human colour matching. *Nature,* 356, 431–433.

Wong-Riley, M.T.T. (1993). Cytochrome oxidase on the human visual cortex. In *Functional Organization of the Human Visual Cortex,* ed. B. Gulyas, D. Ottoson & P.E. Roland, pp. 165–180. Oxford: Pergamon Press.

Yarbus, A.L. (1967). *Eye Movements and Vision.* Plenum Press: New York.

Yamane, S., Kaji, S. & Kwano, K. (1988). What facial features activate face neurons in the inferotemporal cortex of the monkey. *Experimental Brain Research*, 73, 209–214.

Yin, R. (1969). Looking at upside-down faces. *Journal of Experimental Psychology*, 81, 141–145.

Yin, R. (1970). Face recognition by brain injured patients: a dissociable ability. *Neuropsychologia*, 8, 395–402.

Yonas, A. (1984). Reaching as a measure of infant spatial perception. In *Measurement of Audition and Vision in the First Year of Postnatal Life: A Methodological Review*, ed. G. Gottleib & N.A. Krasnegor. Norwood, NJ: Ablex.

Young, M.P. (1992). Objective analysis of the topological organization of the primate cortical visual system. *Nature,* 358, 152–154.

Young, M.P. (1993a). Modules for pattern recognition. *Current Biology,* 3, 44–46.

Young, M.P. (1993b). Turn on, tune in and drop out. *Current Biology*, 4, 51–53.

Young, M.P. (1995). Open questions about the neural mechanisms of visual pattern recognition. In *The Cognitive Neurosciences,* ed. M.S. Gazzanida, pp. 463–474. London: MIT Press.

Young, M.P. & Scannell, J.W. (1993). Analysis and modelling of the mammalian cerebral cortex. In *Experimental and Theoretical Advances in Biological Pattern Formation,* ed. H.G. Othmer, RK. Main & J.D. Murray, pp. 369–384. New York: Plenum Press.

Young, M.P., Scannell, J.W., O'Neill, M.A., Hilgetag, C.G., Burns, G. & Blakemore, C. (1995). Non-metric multidimensional scaling in the analysis of neuroanatomical connection data and the organization of the primate cortical visual system. *Philosophical Transactions of the Royal Society of London, Series B,* 348, 281–308.

Young, M.P. Tanaka, K. and Yamane, S. (1992). On oscillating neuronal responses in the visual cortex of the monkey. *Journal of Neurophysiology,* 67, 1464–1474.

Young, M.P. & Yamane, S. (1992). Sparse population coding of faces in the inferotemporal cortex. *Science*, 256, 1327–1331.

Young, T. (1802). The Bakerian Lecture: on the theory of lights and colours. *Philosophical Transactions of the Royal Society London, Series B,* 92, 12–48.

Zadnik, K., Sataranio, W.A., Mutti, D.O., Sholtz, R.I. & Adama, A.J. (1994). The effects of parental history of myopia on children's eye size. *Journal of the American Medical Association,* 271, 1323–1327.

Zeki, S. (1983). Colour coding in the cerebral cortex: The reaction of cells in monkey visual cortex to wavelengths and colour. *Neuroscience*, 9, 741–756.

Zeki, S. (1990). A century of cerebral achromatopsia. *Brain ,* 113, 1721–1777.

Zeki, S. (1992). The visual image in mind and brain. *Scientific American,* 267, 69–76.

Zeki, S. (1993). A vision of the Brain. Oxford: Blackwell Scientific.

Zeki, S. (1994). The cortical enigma: a reply to Professor Gregory. *Proceedings of the Royal Society of London, Series B,* 257, 243–245.

Zeki, S. & Shipp, S. (1989). Modular connections between areas V2 and V4 of macaque monkey visual cortex. *European Journal of Neuroscience,* 1, 494–506.

Zeki, S., Watson, J.P.G. & Frackowiak, R.S.J. (1993). Going beyond the information given: the relation of illusory visual motion to brain activity. *Proceedings of the Royal Society of London Series B*, 252, 215–222.

Zeki, S., Watson, J.P.G., Luek, C.J., Friston, K., Kennard, C. & Frackowiak, R.S.J. (1991). A direct demonstration of functional specialisation in human visual cortex. *Journal of Neuroscience,* 11, 641–649.

Zihl, J., von Cramon, D. & Mai, N. (1983). Selective disturbance of movement vision after bilateral brain damage. *Brain,* 106, 313–340.

Index